5 STEPS
TO SUCCESS

FOR THE DIGITAL AGE AND BEYOND

Studies in the Application of
Logical-Mathematical Intelligence

Dr. Georg J. Schlueter

Industry Adviser and Educator
Adj. Prof., Aliant International University
San Diego Campus, California

Stellar Literary Press and Media
1968 S. Coast Highway
Laguna Beach, California 92651

Published in the United States of America

ISBN 978-1-962730-25-9 (SC)
ISBN 979-8-89395-915-4 (HC)

Dr. Georg Schlueter Publishing
1480 La Linda Dr.
San Marcos, CA 92078
+1 619. 857. 5019
georg@hansatek.com

Ordering Information and Rights Permission:

Quantity sales. Special discounts might be available on quantity purchases by corporations, associations, and others. For details, contact the publisher at the address above.

For Book Rights Adaptation and other Rights Permission. Call us at toll-free 1-888-945-8513 or send us an email at admin@stellarliterary.com.

Preface

In today's digital-technology-dominated age, business intelligence is not just for the community of business managers; it is for all of us. Digital technologies are affecting the way we live and work. We are all dealing with knowledge management related tasks on a daily basis. We are dealing with data, processing it for coherent information and face the challenge of translating the information into actionable knowledge.

The 3rd Industrial Revolution, which was marked by process re-engineering and knowledge-based process automation, achieved astounding increases in productivity. The ongoing 4th Industrial Revolution, marked as Digital Transformation and supported by Artificial Intelligence (AI), pursues super-effective process mining, digital twin simulation and robotic process automation (RPA). In this 4th Revolution, Business Intelligence (BI) in the form of Logical-Mathematical Intelligence (LMI) plays a major role in shaping newly emerging enterprise architectures (EA) and virtual infrastructures (VI). Industries experience exceptional growth in productivity, which becomes the foundation of a broad range of societal benefits.

The diverse components coming together in the 4th Industrial Revolution form a perfect storm that needs to be harnessed to achieve the promised benefits. For this purpose, the Synergetic Five-Step Innovation Process (SFSIP) was created, which organizes the path forward into a coherent process and enables an assessment of visionary entrepreneurs' ideal conflict-free solutions. Ideal conflict-free solutions portend exceptional resolves. To unleash extraordinary benefits, the envisioned solutions need to be assessed for their functional characteristics and their risk factors.

In our conscious decision-making process, we tend to apply logical intelligence although our conscious mind is far from having the needed intelligence ready for every situation. As a consequence, failures and disasters happened in the past and will happen in the future. We can minimize failures and the potentially catastrophic consequences by applying the novel Synergetic Five-Step Innovation Process.

Systems, whether planned or not, tend to follow their own logic toward successes and failures. Once a failure mode has entered into its vicious cycle, a self-feeding natural instability marks the sequence of events, and there seems to be no way out. Sometimes, though, natural logic does offer a fortunate turn of events, and the failure and its catastrophic outcome is magically prevented. Such magical turns of events are typically referred to as "blessings in disguise." It is this "mysterious disguise" that needs to be unraveled in advance.

When certain aspects of project-inherent Logical-Mathematical Intelligence (LMI) have been overlooked during the planning phase, the risk of failure can outweigh the benefit. Such programs may be characterized as a "wolf in sheep's clothing." Initially, the project may proceed as envisioned. However, during the course of events, the wolf may shake off its sheep's clothing, push the perceived program advantages a-side and force the unpreventable disaster to happen.

The power of the Synergetic Five-Step Innovation Process with Logical-Mathematical Intelligence at its center is unfolded in the analysis of historic innovations and is applied to the demanding needs of modern societies' emerging mass mobility.

The journey through historic innovations reveals a breakthrough observation. Digital technologies lend themselves to the development of intangible virtual infrastructures, which, if superimposed onto physical infrastructures, raise their capacity significantly. Thus, the 4th Industrial Revolution is marked by virtual infrastructures and ends the period of inefficient expansions of physical infrastructures. For the transportation industry as an example, this evolution can finally lead to mobility for all.

Dedication

My thoughts on how productive innovations are created and tragic failures can be prevented shall be dedicated first and foremost to my wife and lifelong partner, Tosca, who has lovingly supported my aspirations for more than 60 years; my late father, Heinrich Schlueter, whose legacy created an inspiring environment during my childhood and continued to affect my ambitions at the university and during my professional years; my late mother, Lisa Schlueter, who enabled an insightful academic education during my adolescent years; our son, Frank, who practices similar thoughts in his successful career as IT director; our daughter-in-law, Nicole K., who contributed to the book's cover design; our daughter, Nicole Luise, for her unconditional love; and our grandchildren, Clair and Kate Schlueter, with the hope that my thoughts can contribute to brighten their future.

Acknowledgements

I want to thank my colleagues at Alliant International University, professors Rene Naert, Cory Scott, and Aaron Wester for their valued discussions and contributions. I owe special thanks to Dr. Rene Naert, a strong promoter of emerging leadership principles in industry and academia, who encouraged me to write down my thoughts on the ongoing integration of business management principles and advanced technologies.

My thanks shall be extended to include business adviser Dr. John McCready for sharing his thoughts on cyber security and evolving standards; Dr. Reza Hosseini for his contribution on digital marketing; and to Frank Schlueter, IT Director for the Glendale Unified School District in Glendale, California, for sharing his thoughts on effective project execution in the digital age.

Contents

Figures and Charts

Abbreviations

ft	Feet
m	Slope of Supply and Demand Lines
mph	Miles per Hour
p	Supply and Demand Line Intercepts with the Price Axis in S & D Diagrams
q	Quantities of, Supply and Demand in S & D Diagrams
vph	Vehicles per Hour, Lane Capacity
vpm	Vehicles per Mile, Traffic Density
ACWP	Actual Cost of Work Performed
AI	Artificial Intelligence
BCWP	Budgeted Cost of Work Performed
BCWS	Budgeted Cost of Work Scheduled
BE	Breakeven (Revenue)
BGWT	Bidirectional Green Waves of Traffic on Bidirectional County Roads
CDMA	Code Division Multiple Access in Wireless Telecommunications
CFD	Computational Fluid Dynamics
CM	Contribution Margin
DCL	Driver Comfort Level, Logical-Mathematical Intelligence of Driving Habits
DT	Digital Twin
DTO	Digital Twin of Organization
EBITDA	Earnings Before Interest, Taxes, Depreciation and Amortization
EP	Earnings Potential = CM = GPM
FC	Fixed Cost per Income Statement
FV	Free-Moving Vehicles
FVS	Free Moving Vehicle Spacing (Center-to-Center Spacing)
GP	Gross Profit per Income Statement
GPM	Gross Profit Margin = CM = EP
GWT	Green Wave of Traffic on Bidirectional County Roads
IoT	Internet of Things
IRR	Internal Rate of Return per MS-EXCEL

JV	Jammed Vehicles in Slow-Moving Traffic Jams
JVS	Jammed Vehicle Spacing (Center-to-Center Spacing)
LMI	Logical-Mathematical Intelligence
LMK	Logical-Mathematical Knowledge
MC	Marginal Cost
MR	Marginal Revenue
NOI	Net Operating Income per Income Statement
NPD	New Product Development
PDMA	Product Development & Management Association
PM	Project Management, Project Manager
Q(e)	Equilibrium Quantity, Supply and Demand Diagram
QFV	Quadratic Function of Velocity
REV	Revenue per Income Statement
ROI	Return on Investment
RPA	Robotic Process Automation
RR	Reserve Rate
S & D	Supply and Demand
SDCTM	Space Division of County Traffic Management
SDFTM	Space Division of Freeway Traffic Management
SFSIP	Synergetic Five Step Innovation Process
TDMA	Time Division Multiple Access in Wireless Telecommunications
VC	Variable Cost per Income Statement

Introduction

The digital transition presents itself as a multi-faceted change. The prime characteristic of this digital transition is that qualitative assessments must be followed with a disciplined quantitative evaluation. The quantitative definition demands a devoted attention more than ever before. The reason for this shift in importance toward quantitative definitions of processes is because of the wide spread need to automate and integrate work processes. Computing devices can only deal with numbers and precise numeric definitions of processes.

The ability to understand logical patterns and the intelligence to organize and quantify objects form the characteristics that distinguish humans from other species. It is in the human nature to constantly search for the meaning of the world around us. [14] The ability to use numbers correctly and the ability to reason are essential elements in today's field of study and in the workplace.

Logical-Mathematical Intelligence (LMI) is one of the eight intelligences that make up Howard Gardner's Theory of Multiple Intelligences. [12] Gardner defines intelligence as the ability to solve problems or produce products that are of importance in a given cultural context or community. Logical-Mathematical Intelligence is the human ability to identify logical or numerical patterns.

LMI is not just practiced consciously by humans. Natural processes follow logical-mathematical patterns. Even the human brain performs logical-mathematical assessments to guide us constantly in an unconscious manner to deal with and respond to external stimuli. [2,17]

Being aware that logical-mathematical intelligence is imbedded in natural events and is being practiced by our brain leads to the conclusion that a conscious knowledge of logical-mathematical intelligence of projects and processes that we want to control becomes an important requirement.

Disciplined and consequential consideration of logical-mathematical intelligence defines a secure path to success. In contrast, when nature's logical-mathematical intelligence has not been fully understood and has not been fully considered, disastrous project and process failures can happen. In those cases, humans were most likely misguided by strong desires that affected their thought processes and disregarded a disciplined assessment of the logical-mathematical intelligence involved.

The purpose of this book is to introduce the Synergetic Five-Step Innovation Process (SFSIP)[1] as a disciplined methodology for addressing the logical-mathematical intelligence underlying successful disciplined innovation processes.

It is about how to envision and design intelligent, safe, and resilient infrastructures for today that will be functional and effective for society in the future. The future will be marked by invisible virtual infrastructures that will remain hidden from our conscious world, despite the fact that they will have far-reaching impacts on international trade, the performance of the world's economies, and how we will work and live. [32]

Virtual infrastructures will greatly affect the economy in the digital age and, thus, will assume a pivotal part of the business intelligence of the future. Their technology-based nature makes them invisible and difficult to be envisioned by less experienced entrepreneurs. They remain invisible when installed. Their technology-based nature emphasizes the importance of LMI as an important methodology in issue-resolution, idea-definition, and solution implementation processes.

The future infrastructures' invisibility is further exacerbated by an inherent natural phenomenon. Unidentified natural features of systems and processes can emerge at critical moments. Depending on the nature of their unidentified dormant features, they can either support the intended positive outcome, or they can suddenly enforce an unavoidable failure with catastrophic consequences. This observation underlines the importance of LMI as a critically important component of program assessments in the digital age.

The author offers a disciplined approach for identifying and following-through on new processes and business innovations. His Synergetic Five-Step Innovation Process (SFSIP) begins with a thought experiment, which resembles a hypothetical theory about cause and effect related to an issue of interest. Thought experiments can be highly unrealistic, but they can also be the precursor of unprecedented solutions. The novel SFSIP methodology creates a disciplined approach for developing useful thought experiments that can guide the entrepreneur along a realistic innovation exercise so that an ideal conflict-free solution can be born out of the thought experiment.

An entrepreneur's thought experiment and the ideal conflict-free solution must be modeled and tested using sound LMI, which can be drawn from the rich contents of mathematical sciences. This disciplined approach provides a framework for the all-important innovation process in this industrial age that is shaped by digital transformations. The thought experiment and its ideal conflict-free solution form the umbrella embracing the entire Synergetic Five-Step Innovation Process.

Several historic innovations and their continued evolutions into the industrial age are being analyzed using the novel SFSIP methodology with its LMI technique in its center. The purpose is to illustrate the dynamics of innovation processes and show that the logical progression of inventions' successes and failures can be disclosed by the SFSIP methodology. The objective is to enhance future innovations and to prevent failures.

This book is not meant to present a complete history of innovations but rather to discuss several phenomenal episodes that illustrate the uniqueness of key historic innovations, hidden features, and their lasting impacts on mankind. The detailed outlines of lengthy innovation processes combined with a discussion of emerging technologies serve as illustrations of the dynamics and the speed at which they emerged and the accelerated speed at which they need to be created and tested in a meticulous manner. Nothing can be left to chance anymore.

In the context of advanced LMI in the digital age and the SFSIP, Columbus' courageous discovery of the Americas, the sinking of the unsinkable Titanic, the great worldwide depression of 1929/30, and the nuclear accident at Chernobyl are viewed in a completely different light.

The problem-solving power of the Synergetic Five-Step Innovation Process with the Logical-Mathematical Intelligence in its center is illustrated in Chapter 11. The worldwide issue posed by traffic congestions on freeways and county roadways is addressed, and ideal conflict-free solutions are introduced. To gain actionable insight into the complex dynamics of traffic flow and its worsening trend of forming slow-moving traffic jams, intelligence is developed in the form of numeric definitions and breakthrough solutions.

Chapter 1
Creation and Assessment of Innovations

Introduction

Resources to be considered for raising competitive advantage include human ingenuity more than ever before. Human ingenuity must be engaged as an umbrella resource to expand the boundaries of traditional resources and open new avenues toward continued growth in productivity on all economic fronts.

In the digital age, perpetual innovation and global competition made the traditional organizational architecture obsolete. Modern organizations clearly favor flat structures, where the new knowledge workers can freely flourish within e-business-based environments. In this new form of organization, groups of workers decide for themselves what to do and are held accountable for the outcome. [24,25,51]

The author offers a disciplined approach for identifying and following through on new processes and business innovations by following his guiding Synergetic Five-Step Innovation Process. [45] The five-step process begins with a thought experiment, which resembles the vision of the entrepreneur and his ultimate objective of his innovation exercise. The thought experiment must be modeled and tested using Logical-Mathematical Intelligence, which can be drawn from the rich contents of mathematical sciences.

Synergetic Five-Step Innovation Process (SFSIP)[2]

Business development involving new products and services has been structured into qualitative and quantitative research. Qualitative research is used to gain a general sense of phenomena that cause issues and to form concepts for their explanations. Quantitative research collects quantitative data and is expected to translate abstract concepts into quantifiable measures. Quantitative research is expected to systematically examine whether there are cause-and-effect relationships between variables. Statistical methodologies are being used in many instances to interpret the data and answer specific questions.

Qualitative and quantitative research are important and can be complementary. Both research modes work essentially hand-in-hand with the right statistical approach, though, today, we need to work harder on statistical analysis. Statistics typically deliver trends and cannot necessarily provide the specific information needed for a practical solution to a problem. Explicit mathematical correlations are needed in today's digital age in order to correctly assess the situation and find ideal conflict-free solutions.

The author modified the traditional qualitative and quantitative research to create a methodology for mastering the challenges in today's age. He promotes his SFSIP, which entails the creation of a thought experiment as part of the qualitative research phase and suggests LMI as part of the quantitative research phase. The thought experiment shall yield a disciplined focus on practical results, and the LMI shall facilitate a departure from traditional statistical representations of complex concepts. Logical-Mathematical Intelligence can yield more tangible definitions of complex systems.

The thought experiment is synonymous with traditional qualitative research. There is a difference between the two although both really mean similar things. As we try to combine the two, we find that it is useful to consider both separately and define both as separate sub-components of the first step of the innovation process.

Step 1, Qualitative Research

Qualitative Research defines the framework of a selected issue or problem. This is typically accomplished by research of existing literature to identify existing information on the subject matter and to explore conclusions offered by earlier researchers. [14] In today's age, with the Internet and powerful search engines, a huge volume of non-numerical information is readily available on any subject. When the issue being

6

addressed is indeed an ongoing problem for a business entity, an entire industry or the society as a whole, the researcher will most likely observe that the existing information is incomplete and the prevailing findings are inconclusive. This leads to Step 2, the innovative imagination of an ideal, conflict-free solution.

Step 2, Ideal and Conflict-Free Solution

As is illustrated in Figure 1.1, once the qualitative research has resulted in a saturation of information for the researcher, an ideal and conflict-free solution can be born out of the entrepreneur's creative intuition and his power of imagination. [21] This part is one of the new concepts in research in today's digital age. The perceived solution can represent a breakthrough concept that guides subsequent efforts. It defines the path forward and becomes the object of detailed quantitative evaluations.

Step 3, Logical-Mathematical Intelligence (LMI)

After having identified an ideal and conflict-free solution, the quantitative research can commence, which is the process of collecting and analyzing numerical data.

In this phase of a research project, when researchers are faced with an overwhelming complexity of the project and its big data, they tend to escape into statistical methodologies, with the result that the knowledge gained is not transparent and remains vague. [13] In contrast, disciplined Logical-Mathematical Intelligence provides the ability to reason and facilitate the disclosure of logical and numerical relationships. LMI facilitates critical thinking to solve challenging problems with the objective to identify one or more correlations that defines the thought experiment in numerical or algebraic terms.

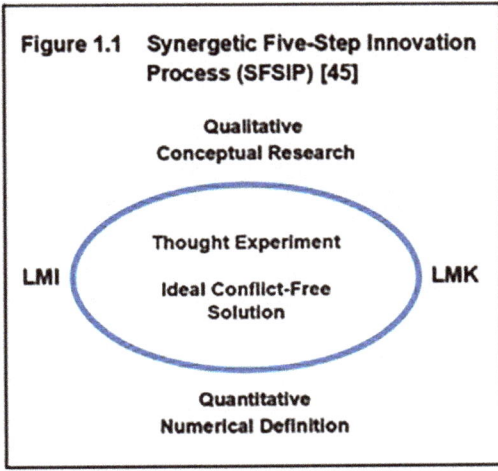

Figure 1.1   Synergetic Five-Step Innovation Process (SFSIP) [45]

Qualitative Conceptual Research

Thought Experiment

LMI

Ideal Conflict-Free Solution

LMK

Quantitative Numerical Definition

7

Step 4, Logical-Mathematical Knowledge

Exercising the numerical and/or algebraic terms to examine the perceived solution critically from various angles yields Logistic-Mathematical Knowledge (LMK) about the problem at hand and about the perceived solution. LMK enables the researcher to engage in a true iteration process to adjust and upgrade the initially perceived ideal solution. In the process, the originally perceived ideal solution will either be confirmed or modified and adjusted as needed to design a practically feasible version.

Step 5, Detailed Solution Assessment

In the final step, a detailed analysis is conducted to assess the benefits and the risks offered by the ideal conflict-free solution. If the quantitative research lacks logical mathematical intelligence and applies non-explicit analysis, a complex system will remain inadequately defined. In contrast, standardized data collection protocols and tangible definitions of abstract concepts enable replication and, thus, provide sound confirmation of the hypothesis.

In summary, the Synergetic Five-Step Innovation Process reaches deeper and further than traditional optimizations. An optimization is an approach where the basic process remains unchanged and where the least adverse procedure is identified and implemented. This traditional approach oftentimes overlooks the opportunity for completely different conflict-free solutions. Rich opportunities are particularly offered by the currently emerging digital technologies, such as artificial intelligence (AI), the Internet of Things (IoT), process mining, digital twin applications, machine learning and high-speed computing technologies. Considering these technologies in combination with the SFSIP can yield ideal conflict-free solutions with unprecedented benefits for the industry and the society as a whole.

Logical-Mathematical Intelligence (LMI) and Knowledge (LMK)

There is a trend to simplify complex systems. The abundance of variables of complex systems makes people search for the governing variables and overlook the influence of seemingly less important parameters. Einstein said once, "You must simplify," and then he warns, "But do not over simplify." [23] In essence, one has to simplify, but in the simplification process, one must consider all variables. That's where the challenge of meaningful simplifications lies. The secondary parameters of complex systems need to be considered as well because they may disclose a system's dark side and its inherent instability.

Critical thinking is an exercise that is meant to guide people through lengthy and complex thought experiments to prevent shortcuts. All parameters and all associated system functionalities must be included in the analysis.

In the digital age, perpetual innovation and global competition have made the traditional organizational architecture obsolete. Modern organizations clearly favor flat structures, where the new knowledge workers can freely flourish within e-business-based environments. In the new form of organizations, groups of workers decide for themselves what to do, and they are accountable for the outcome. The old command and control structures tend to impede the spread of knowledge and limit the economies of scale that could be reaped. In contrast, the new flat organizations encourage creativity and innovation. [28] Thus, innovation can be quite a simple thing. It does not reside anymore in the minds of brilliant scientists. It can blossom almost anywhere in organizations that are properly structured and provide a culture that makes invention flourish. Consequently, it is critically important to spread the concept of Logical-Mathematical Intelligence and Knowledge to the groups that are expected to foster innovations.

Examples of fine successes and catastrophic failures are plentiful throughout history. Most stories of beautiful accomplishments can be traced back to their underlying thought experiment and the successful application of LMI. Likewise, stories of tragic accidents can be traced back to an incomplete thought experiment and the omission of LMI. The five-step approach in the analysis of failures not only reveals the omission of logical-mathematical intelligence but, can also reveal the falsehood in the original thought experiment. Unfortunately, as long as the falsehood in the thought experiment has not been fully identified, society will tend to make the same mistakes again and again, albeit each time under different pretentions. As a result, society must endure the tragic consequences. The tragic repetitions of failures of the same type occur not only in technical fields, but in businesses, economics, and political environments.

Progress occurs in spurts and is driven by impactful innovations. Progress can also be halted by failures and their accommodating disasters. We must be aware of these two phenomena to enforce the one and avoid the other.

The number of programs and applications that can be utilized to define functional structures of the thought experiment and to exercise the logical-mathematical knowledge is rising in the digital age. The systems' analytical capacities are greatly expanded by AI technologies. While AI-driven technologies enhance the analytical capacity tremendously, the key aspect of the task remains to be the upfront ideation process and its expansion toward a complete and meaningful thought experiment, which remains the main task of the inventor.

The economic system with its various components can be treated as a logical thought experiment. Logical-mathematical intelligence can be applied to explore how the various components affect the course of economic events. Applying the novel SFSIP methodology yields exceptional transparency to show policymakers, the general public, corporate executives, and business management students where the priority lies and how economic derailments can be identified before they happen.

This book offers a disciplined approach for identifying and following through on new processes and business opportunities. The creative mind must adopt a logical path that starts with ideation, expands it into a comprehensive thought experiment, and analyzes it using a critical mathematical assessment to explore its practical feasibility and its intrinsic business and/or societal value. This five-step approach creates a rich productivity-enhancing environment that delivers an abundance of opportunities for successful projects while preventing failure prone innovation processes.

Chapter 2

Historic Emergence of Intelligence and Knowledge

Introduction

An attempt is made to assess several key historic innovation events and their continued evolution into the industrial age. The purpose is to illustrate the dynamics of innovation processes and that innovation successes and failures can be traced back to the SFSIP. The objective is to enhance future innovations and to prevent failures.

This book is not meant to present a complete history of innovations but rather to discuss several phenomenal episodes that illustrate the uniqueness of key historic innovations, hidden fortunate features, and their lasting impacts on mankind. The detailed outlines of lengthy innovation processes combined with a discussion of emerging technologies shall serve as illustrations of the dynamics and the speed at which future innovations have to be created in their completeness.

The Emergence of the Knowledge of Wheels and Ball Bearings

The Wheel is definitely one of the most impactful mechanical inventions accomplished of all time (3,000 BC). It was instrumental in enabling primitive physical transportation in ancient times and continued enabling the industrial revolutions. The industrial era is unthinkable without the wheel. And it is not just the wheel by itself that enabled the industrialization; its derivatives, like ball and rod

Figure 2.1 Ancient Wheel

Internet

11

Figure 2.2   Egyptian Pyramids

Internet

bearings, are equally important. Large and small household appliances, and small and large industrial equipment with rotating components are unthinkable without high-tech bearings.

Cave drawings reveal that the wheel concept inspired the Egyptian pyramid builders to very intuitive applications. The wheel concept yielded enormous improvements in efficiency.

Although the wheel concept and its derivatives evolved over several millennia and seem to have emerged by chance, the development can be traced back by the Synergetic Five-Step Innovation Process. The elements of the SFSIP process are clearly visible if not knowingly enforced by people but rather as a natural law, where people intuitively implemented improvements that offered themselves.

During the pyramid construction period (2,700 -1,700 BC), the Egyptians initially loaded the heavy boulders on sleds and pulled the sleds

Figure 2.3 72 Workers Pull a Granite-Loaded Sled Over Raw Desert Sand   Internet

over bare desert sand. Water was apparently used to ease the resistance, as is indicated by ancient cave drawings shown in Figure 2.3. As shown in the drawing, approximately 72 people were needed to move sleds loaded with multi-ton granite boulders over desert sand.

Later in the period, the Egyptians employed the benefits of rollers, which is an early primitive version of the wheel concept. With rollers placed under a sled, the number of people needed to pull the sled could be drastically reduced. Rollers being placed between the two otherwise sliding

12

surfaces replace the surface-to-surface frictional resistance by a much lower rolling resistance. Under ideal conditions, rolling resistance can be near zero. This is not the case for rolling imperfect rods over desert sand, but the rolling resistance is definitely much less than the surface-to-surface frictional resistance. As the assembly is moving forward, the load is actually gaining compared to the rollers. Thus, the rollers

Figure 2.4   Granite Block Supported by Rollers

Internet

drop out of the system at the back end and have to be carried forward (at twice the speed of the assembly itself) and be placed again at the front end. This is the result of an inherent physical feature of rollers.

Rollers behave like wheels shown in Figure 2.5. As a wheel moves forward at a certain velocity (V), that forward velocity applies to the center of the wheel, while the bottom part of the wheel is firmly resting on the ground

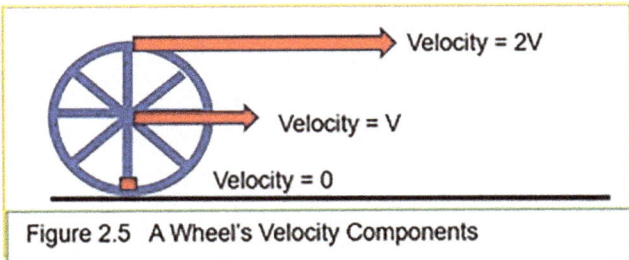

Figure 2.5   A Wheel's Velocity Components

and is not moving forward at all. In contrast, the top of the wheel is moving forward at twice the velocity of the wheel's center velocity. Thus, the great advantage of the Egyptian roller design is that the heavy load, which is resting on top of the rollers, is actually moving forward at twice the velocity of the rollers. The gain of the roller system compared to the basic sliding sled is that the surface-to-surface sliding resistance is replaced by a much lower rolling resistance and that the load is moving at twice the velocity.

Since the workload is moving at twice the velocity of the rollers, the benefits of the roller system can be characterized by the astounding analogy that the distance to be mastered by the rollers over desert sand amounts to half-the-distance of the workload's journey. This is an exceptional benefit of the roller system compared with the future invention of the wagon system, where the load rests on the wheels' axle and moves forward at the velocity of the wheels. The wagon system is, of course, more practical for paved roads. However, dealing with unpaved raw desert sand, the primitive roller system had an extraordinary advantage over the future's more advanced wagon system.

The underlying thought experiment springs out of the idea that the resistance and the transportation time had to be reduced. Making the desert sand wet caused a certain yet limited improvement for the sled-based system. But when the knowledge of wheels emerged and reached the Egyptian pyramid builders, they applied it in the intuitive yet primitive form

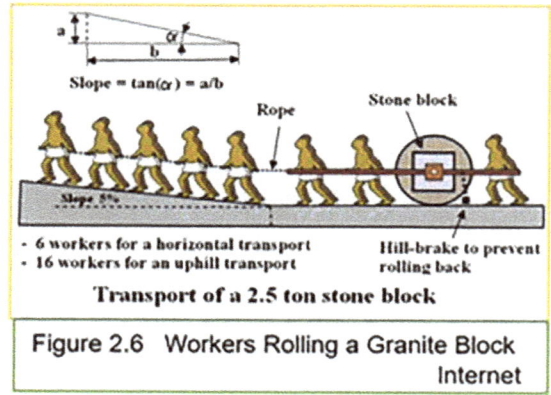

Figure 2.6 Workers Rolling a Granite Block
Internet

of rods and gained a 12-fold improvement in terms of lower resistance, twice the velocity, and half the distance for the rollers. Instead of 72 workers, only 6 workers were needed, which could accomplish the job in less time, an extraordinary increase in productivity.

We cannot say that the technology applied by the Egyptians during the 1,000-year pyramid-building period was standing still. There were enormous improvements, except that the improvements occurred over extremely long periods of time.

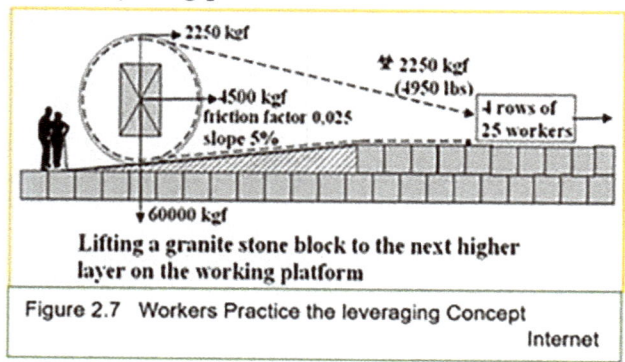

Figure 2.7 Workers Practice the leveraging Concept
Internet

There is evidence that they designed wheel-shaped wooden cradles and placed the heavy granite boulders in the center, as shown in Figure 2.6. Cave drawings reveal that just 6 people were needed to roll those wooden cradles across desert sand, and 18 people were needed to roll the cradle uphill along small inclines.

14

It appears that the Egyptians learned to leverage the features of the wheel in very unusual yet unique ways. They understood the concept of leveraging. They built large wheel-type wooden cradles with boulders in the center, laid a rope around the wheel as shown in the schematic drawing in Figure 2.7, and they had workers pull on the upper end of the rope. A 60-ton granite boulder would require a pull force equivalent to 2¼ ton to pull it uphill along a 5-degree incline. This represents a leverage factor equivalent to approximately 25, an impressive leverage at the time.

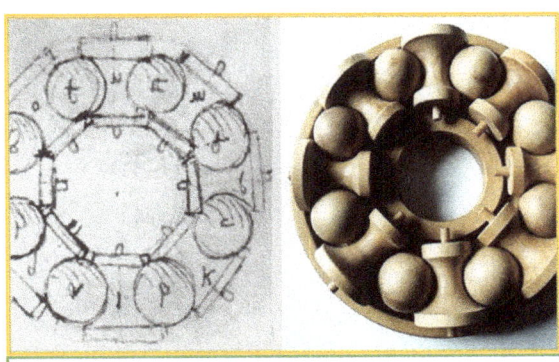

Figure 2.8 Leonardo da Vinci's Ball Bearing Design   Internet

In applications of the wheel as wagon wheel with an axle in the wheel's center, the frictional resistance is transferred from the ground to the wheel's center, where the wheel is rotating around the axle. The improvement accomplished is proportional to the ratio of the wheel's diameter to the diameter of the central hole. This can amount to an enormous improvement for ancient transportation but not enough for the industrial age.

The diverse inventor Leonardo da Vinci (1452–1519 AD) designed several gadgets involving drive shafts and wheels and made the first attempt to design a mechanism that could further reduce the remaining surface-to-surface friction and to extend the durability of the wheel-to-axle interface. He made the ingenious step to apply the wheel concept in the form of spheres although the flat surface to flat surface interface is still present in his initial concept. Shown in Figure 2.8, da Vinci's practical application in form of a wooden ball bearing design

Figure 2.9 Modern Ball Bearing   Internet

15

represents the beginning of the industrial ball and rod bearings as we know them today.

The development of the modern ball and rod bearings had to wait until more durable metallic materials became available in the industrial age. Ball and rod bearings (Figure 2.9) have become essential for all equipment that involves rotating wheels, discs, and rods.

In addition, in the industrial age, the Egyptian roller concept was further advanced and led to the creation of tracked or caterpillar vehicles, which are equipped with continuous bands of steel plates as shown in Figure 2.10. The steel plates underneath the wheels are preventing the wheels to sink into soft grounds and enables the vehicle to climb steep inclines. The same plates are automatically picked up at the back end of the vehicle and are carried forward at twice the velocity of the vehicle and are placed again underneath the vehicle's front wheels. In a way, the modern caterpillar vehicle design is comparable to the Egyptian roller technique except that the modern design replaces the ancient rollers by actual wheels with axles in the center and adds the continuous band of steel plates.

The history of the wheel and the related innovations span a period of 5,000 years. The emergence of modern wheel designs and applications was carried by approximately 170 generations. Yet every link in the chain of innovations can be recognized as being guided by a thought experiment and a logical-mathematical assessment of its effectiveness. This

Figure 2.10 Modern Caterpillar Vehicle with Steel Plates   Internet

thought experiment most likely existed in the minds if the leaders of the time. The mathematical assessment of their thought experiment may not have been carried out the way our LMI dictates us to proceed, but LMI was present at the time like natural laws and acted unbeknown to the

16

people at the time. The beneficial effects of the thought experiment were strikingly obvious as the goals were simply a reduction in the number of workers and the time needed for the tasks.

It is interesting to note that the wheel by itself can be viewed as an automated roller system. While the rollers had to be carried forward at essentially twice the speed to catch up with the front of the object, the wheel's top also moves forward at twice the speed of the object although that movement is an inherent feature of the wheel and happens automatically

We will see that the visibility of the projects of the antique gradually transitions toward less visibility and more intangible objectives. Thus, the awareness of structured innovation processes consisting of a well-defined logical thought experiment and a disciplined application of logical-mathematical intelligence becomes increasingly important in a world where innovations involve intangible subject matters.

The Impact of Greek Philosophers on Today's Logical-Mathematical Intelligence

The development of LMI can be traced back to the Classical Period of the Greek culture. Logic and the skill of critical thinking were initially exercised in disciplined manners by Greek philosophers (600–300 BC). They followed through on their quest for practical information with critical analyses using LMI wherever possible. If it did not exist yet, they researched it and established it. They established LMI and created knowledge, making it available for others to apply it toward innovative improvements of the way things were being done. Their intellectual works and legacy influenced the course of Western Societies. We should briefly mention the four most influential philosophers who learned from each other and created the classical period of the antique.

Pythagoras (570–495 BC) was interested in mathematics although it was not allowed by elite circles in Greece at the time. They wanted to judge by a perceived beauty. A group of elders, the oracle, was queried about architectural and political, social and war related issues. Answers predicting the future were typically framed in an ambiguous manner so

17

that its meaning depended on interpretation and, thus, could never be proven wrong. The oracle's answers were widely respected because the intellect at the time was strongly affected by religious thinking, myth, and a strong perception of beauty as a divine ruling. The configuration of a new temple was not pre-designed by applying proportions related to an existing beautiful temple. A new temple had to be configured using the oracle's perception of beauty. Consequently, there was no room for Pythagoras' logical experiment for developing rules of proportion.

Their thought experiment gave Pythagoras and his followers a strong conviction and the courage to dedicate their lives to their dreams. Pythagoras and his scholars went to Southern Italy where they formed a brotherhood and practiced their mathematical endeavors secretly within a fenced-in enclave. The circumstances of various discoveries are not exactly clear, but Pythagoras was credited with several lasting discoveries, including the Pythagorean Theorem, the Triple, and the general concept of proportions. We will discuss these discoveries in greater detail.

The lives of Socrates (470–399 BC) and Pythagoras were 25 years apart and there seems to have been no intellectual connection. Socrates, who lived and worked in Athens, established the concept of knowledge building through dialog involving questions and answers. The Socratic method was documented by his student Plato (427–347 BC) who continued his work.

Aristotle (384–322 BC) was a student of Plato's. Their lives overlapped for 37 years. Aristotle's work covered a broad range of subjects; most notably, he wrote about natural sciences and economics. He expanded on Socrates' and Plato's ideas on knowledge development and knowledge transfer through teachings. Later in his life, he tutored a young aristocrat named Alexander, who grew up to become Alexander the Great (356–322 BC). Inspired by his excellent education, by the age of thirty, Alexander had created one of the largest empires in history.

Alexander's education in natural and political sciences obtained from his tutor Aristotle came to unprecedented fruition. Aristotle planted in Alexander a logical thought experiment involving education for the world. As a result, one of Alexander's creations was the city of Alexandria in Egypt, where he transformed his teacher's main thrust, i.e., education, into a symbolizing reality. Alexandria became the world's knowledge capital.

Aristotle's and Alexander's spirit of knowledge and education was carried on by Alexander's successors and resulted in one of the world's largest central libraries in the city of Alexandria shown in Figure 2.11. That library became the decisive collection of knowledge and was visited by the world's leading scientists, including, for example, the famous Euclid, a Greek logician, a geometry and number theorist who lived in Alexandria and worked at the library (about 300 BC). Euclidean geometry, which defines the inherent metric laws of triangles, pyramids and spheres, remains critically important in today's high-tech time.

Figure 2.11   The Library of Alexandria (283 BC)
Internet

The Greek mathematician and physicist Archimedes (287–212 BC) studied at the Library of Alexandria and had access to the library's knowledge at his time. One of his discoveries, known as the Archimedes Principle (264 BC), defines the buoyancy that a body experiences when it is placed into a fluid. Archimedes discovered that buoyancy is equivalent to the weight of the displaced water and that it counteracts the weight of the body. Thus, a body floating on a fluid has reached a balance between the two forces.

The Central Library of Alexandria enabled an acceleration of inventions. Alexander's one-on-one education with Aristotle experienced a tremendous multiplication reaching every science-oriented individual in Europe for many centuries. Alexander the Great and his successors were not just military experts and powerful conquerors; they were also extremely effective in carrying on the Greek philosophers' work in terms of knowledge creation and its effective spread.

Using today's knowledge of decisive principles in natural and social sciences and looking back to the time of the Central Library of Alexandria, one recognizes that the library represents the focal point of knowledge collection and distribution. This principle resembles a hub-and-spoke system, which is associated with very effective LMI for analyzing and calculating the efficiency of the system initiated by the original logical thought experiment.

The logical-mathematical intelligence associated with the hub-and-spoke system compared to less disciplined systems shall be discussed later in great detail.

The Inventions of the Hanseatic League

The beginning of the Hanseatic League (the Hanse), rests in northern Germany with trade centers in Lübeck, Hamburg, and Bremen. [18,50]

The Hanse, a trade league embracing Western and Northern European countries as shown in Figure 2.12, evolved over a long period, during the 11th, and 12th centuries. It reached its peak in the 13th and 14th centuries and endured into the 16th century. The Hanse, because of its international reach in northern Europe and its economic power, exercised significant influence on cultural and religious movements and, thus, is a major factor in Europe's historic development.

During the Hanse, German craftsmanship started to blossom, and merchants started traveling from city to city offering the products of their home cities to residents in neighboring cities.

Figure 2.12 The Hanseatic League Practicing International Trade Across Europe   Internet

20

During its heyday, its membership counted more than 400 cities reaching from Aachen, Germany in the west to Königsberg in the east, and in the north, from Bergen, Norway through Denmark and Sweden, all the way to Turku, Finland. The majority of the cities were located on the European continent and engaged in east–west trades, where exotic materials of the east were made available to the skilled craftsmen of the west who converted the unique raw materials into valuable products.

Figure 2.13   The Hansa Cog, A Flat-Bottom Commercial Sailing Vessel   Internet

The initial land-based commerce on the continent became the backbone of sea-based trade across the North Sea and the Baltic Sea. A different set of natural resources, including foods, was made available to the vast number of cities in Continental Europe. [18]

The Hanse established locally managed business centers in London, England and Bergen, Norway to monitor the commerce across the North Sea. Likewise, the Hanse founded locally managed business centers around the Baltic Sea to monitor the commerce interests there. As a result, the development of well-functioning transportation means on land and water flourished. For example, the Hanse developed sea-worthy vessels, the Hansa Cog [19,50], which featured great stability and, because of their flat-bottom design, could stand upright on Wadden Sea at low tide. (Figure 2.13)

Historians interested in the Hanse, its extraordinary economic success, and its enormous influence on cultural, religious, and political developments, tend to focus on the Hanse's impactful societal events. They focus on trade relationships and dissect their impact on the various

21

political and religious powers and successfully illustrate the dynamics involved. [4,50]

Since every successful business engagement is typically born by a logical thought experiment, we need to search for that initial thought experiment that enabled the unprecedented success in trade that led to the formation of the Hanse, where the initial success was multiplied many-fold over a period of more than 400 years.

The literature refers to initial "local" trade activities where product-selling merchants traveled from city to city and that this localized activity evolved into long-distance commerce. For the inquisitive entrepreneur, the question arises how this transition may have been enabled and fostered.

Initially, the formation and growth of cities became a driving force

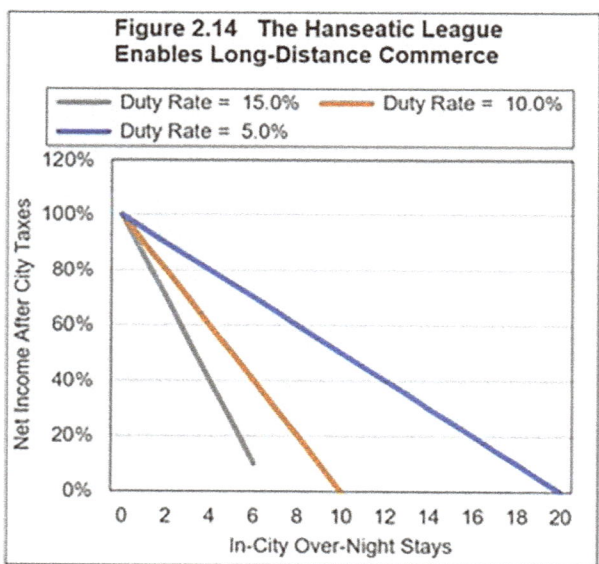

Figure 2.14 The Hanseatic League Enables Long-Distance Commerce

for trade. Salespeople traveled from city to city and offered their merchandise for sale. Cities had surrounded themselves with strong murals to be protected against organized attacks and occasional robbers. Traveling salespeople with valuable merchandise preferred to spend the night inside the cities' murals to have their merchandise protected. The city fathers saw the opportunity to generate extra income and charged the traveling merchants a duty fee for the merchandise in their possession. As shown in the chart, such duty fees adversely affected a travelling merchant's economic rewards and narrowed his activity region severely, which is illustrated in Figure 2.14. For example, at a 10% duty fee, the resulting economics limited the number of intramural overnight stays to less than 10 cities.

Figure 2.15 shows the maximum number of possible in-city overnight stays as a function of the cities' duty rates. The formula for that functionality is

In-City Over-Night Stays = 1 / Duty Rate

This is a typical One-Over-X function that approaches infinity as the duty rate approaches zero. Infinity implies here that the salesmen's travel range would be unlimited. The formula and the chart are a mathematical proof of the brain's inherent logical decision process.

The Hansa city fathers may not have had the benefit of the mathematical proof, but they recognized the positive economic impact of a zero-duty-rate, which became one of the mutually agreed arrangements of member cities.

To accommodate zero-duty overnight stays within city murals, Hansa members provided a fenced-in area inside their city murals, where traveling merchants could stay overnight without having to pay taxes for the merchandise in their possession. This arrangement was not just applied by cities for land travelling merchants but was also applied to the harbors along the Hanse's Sea passages throughout the Baltic Sea. The duty-free zone in harbors and cities

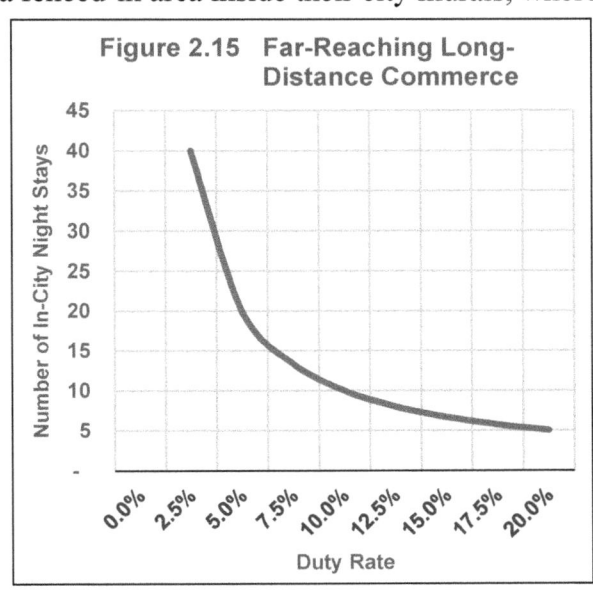

Figure 2.15 Far-Reaching Long-Distance Commerce

became a significant feature contributing tremendously to the Hanse's multi-century long economic success.

The Logical-Mathematical Intelligence of the Hanse's duty-free zone survived the Hanse and has been applied to national and especially international trade ever since. Duty-free zones exist at harbors and airports around the globe and form a substantial backbone of modern international trade.

The Hansa League's impact on the wealth of major member cities, such as Hamburg, Lübeck, Köln, and Rostock was extraordinary and is still visible in those cities' general makeup today.

The Invention of Christopher Columbus

Christopher Columbus (1451–1506 AD) was a gifted entrepreneur in his own ways. To appreciate Columbus' exceptional accomplishment, we need to look at the navigational techniques of his time and his extraordinary entrepreneurial contribution that would enable open-ocean navigation.

For navigational purposes, astronomical data had been assembled into tables to support latitudinal determination (north–south orientation). Although the longitudinal determination (east-west orientation) is equally important, no reliable method had been established yet at Columbus' time. Columbus' entrepreneurial spirit, combined with his extensive seafaring experience, resulted in a pioneering vision that provided him with the missing information to determine his east-west locations along his journey across the open ocean.

The concept is based on the knowledge that one full rotation of the Earth takes 24 hours and that the Earth's circumference at the equator is 24,000 miles. This translates to 1,000 miles along the equator per one-hour time difference. Of course, at north and south latitudes,

Figure 2.16 Columbus' Cross Atlantic Journey Shown on a World Map

Internet

the circles around the globes are shorter than at the equator. The daily time

24

in every region around the globe is considered the moment at which the sun has reached its highest position on the sky. The local time is set at 12 noon. Consequently, if one has a chronometer on board ship and reads the time when the sun has reached its highest position, one can determine the time difference compared to the time at the homeport and translate the time difference into miles.

All clocks in Columbus' Europe were driven by descending weights and were controlled by precisely timed swings of a pendulum. Pendulum clocks require a steady base and, thus, are not useful on rolling ships. An Italian architect had picked up the idea that a wound-up thin metallic stripe can act as a spring and drive a mechanical system, including a mechanical clock. The first spring driven clock is reported in 1410. [4]

By the time Columbus was approximately 30 years old, he had developed his clock-based navigational strategy that allowed him to navigate his way on the open Atlantic Ocean. At the same time, Europe's lucrative spice trade with India blossomed. People practiced an overland route and a sea-passage along the African west coast and around the Cape of Good Hope. Both journeys were plagued with severe difficulties and financial losses. Columbus had the idea to explore a westward sea passage to India. The idea to create a competitive and less cumbersome sea passage to India gained momentum.

Figure 2.17 Typical Commercial Sailing Vessel at Columbus' Time   (Internet)

Columbus, because of his advanced navigational methodology, was convinced that he could be successful in establishing the westward sea passage to India. In the 1480s, Columbus pursued a persistent lobbying effort to several Catholic monarchs in Portugal and Spain and made several proposals. His initial proposals were denied because his estimated distance of 2,400 miles was felt to be only one-quarter of the monarchs' estimates. Eventually, he was able to establish a contract that specified a small amount of upfront funding, a success premium, and 10% royalties in future trades. Years later, after Columbus' death, it was revealed that his sponsors did not believe that he would ever return.

In the fall of 1492, Columbus set sail across the Atlantic Ocean with three ships like the one shown in the picture. After sailing for approximately 4,000 miles, he reached the San Salvador islands. He returned safely and repeated the same journey three more times. His logbook reveals that each time he traveled along the same route and reached the same group of islands, which proves his navigational skills.

It should be noted that Columbus, in addition to his astronomical, geographical, mathematical, and navigational knowledge, possessed the people skills required to execute his pioneering entrepreneurial vision. During Columbus' time, the ancient definitions of the terms "Sea" and "Ocean" were still very valid. While a sea was understood to be a body of water surrounded by coastlines and beaches, like the Mediterranean and the Baltic Seas, the term "ocean" referred to open waters with no coast or beach surrounding it. Still, Columbus was able to convince enough sailors to sign up as crew for all three ships of his first adventure. Maybe his low distance estimate of the journey was part of his strategy to hire the crew he needed. It is known that he issued false travel advances to his crew in order not to scare them when considering the distance from home. Otherwise, he correctly measured his sailed distance over ground and was able to correctly determine the 4,000-mile distance to the San Salvador Islands. However, later assessments of his landing site indicate that he actually may have landed at the Haitian islands, which are exactly 4,000 miles away from Portugal.

The astronomical data assembled for his smaller Earth indicated that for his landing site there was a certain eclipse visible in the sky. Columbus found that the stars in question were not lined up as predicted. But he did not bother to research that circumstance any further. He was a practitioner, a seafarer, and not a research-oriented scientist, and he may not have been familiar with the rules and laws of Euclidean geometry. If he had done the geometric analysis to explain the deviation from the perfect eclipse at his location at the time, he could have provided an exceptional proof for the more accurate size of the Earth.

Columbus claimed throughout his remaining life, that he had reached India, a claim for which he is openly criticized for even today. However, one must remember that the main rewards were not paid up-front but were conditioned on his success. Thus, he had to stick with his claim to qualify for the promised reward.

Columbus is undoubtedly credited for the discovery of America, and the story presented above is consistent with typical descriptions in honor of his pioneering accomplishments. His discovery that time and distance around the globe are interconnected and his pioneering move of adding the clock to his navigational tool set are simply remarkable. His extremely low estimate of the westward distance from Portugal to India, his accurate determination of the actual distance travelled to the San Salvador or Haitian Islands, and his persistent claim to have reached India, all remain a mystery that deserves further investigation.

Combining the mysterious observations into a set of logical-mathematical intelligence and conducting a more detailed assessment provides interesting conclusions, which add to the enormous strength of Columbus' spirit as a seafaring explorer. The assessment is described in Chapter 9.

Chapter 3
Logical-Mathematical Intelligence in Social Sciences

Introduction

Today's economy offers overwhelming choices for everyone and everything, a truly global marketplace with extremely intense competition and aggressively reduced prices. Business owners are continuously trying to cope with critically reduced margins by attempting to reduce their costs. This puts a heavy burden on today's innovative improvements in productivity and service quality.

While the number of choices offered in the marketplace is expanding and seems to become unlimited, the limitations of resources become more apparent and tend to drive the costs of all resources upward. The latter makes it very clear that the resources to be considered for raising competitive advantage must include human ingenuity more than ever before. Human ingenuity is engaged as an umbrella resource to expand the boundaries of traditional resources and open new avenues toward continued growth in productivity on all economic fronts. It has been said that natural resources must not be consumed anymore; rather, they shall be borrowed and be used as catalysts in production processes. These are not simple transitions and call for deeply rooted innovations.

This trend is further acerbated by the need to raise environmental compliance across all industries. The primary concern is air pollution and the resulting climate change, which calls for drastic changes in resources and energy related strategies. The change must occur in harmony within well-functioning economic frameworks, i.e., without catastrophic economic downturns. The economy obeys its own natural rules, and industrial restructuring must be accomplished without violating those rules.

In their attempt to accomplish steady growth of their economies, modern societies experience a continuous struggle between price inflation, wage growth, and unemployment. Each of these components of the economic system are clearly visible to watchful observers and, thus, attract the society's attention, although none of these components can be affected directly by the government's policymakers. Fortunately, there is another component that has the power to directly counter-affect annual inflation and unemployment while allowing reasonable annual wage growth. That controlling effect is achieved by productivity growth. Productivity growth is the most powerful economic stabilizer. It is a solid source of wealth creation in modern societies. Unfortunately, productivity growth is an intangible and invisible component of the economic system and, thus, it is not in the mind and not in the immediate interest of the general public. This circumstance makes it difficult for elected policymakers to put productivity and productivity growth at the very top of their list of priorities, but they should. This takes us directly to methodologies that can strengthen the path to ingenuity, innovation, and productivity growth.

The Inventions of Jakob Fugger

Jakob Fugger (1459–1525 AD) was a major German merchant, mining entrepreneur, and banker. His enterprises primarily centered on the lending of money to merchants and aristocrats. It appears that Jakob Fugger learned about double entry bookkeeping, banking, commercial law, and financial mathematics during the years 1473 through 1487, when he resided at a very young age in the Venetian Republic. He also learned all about the financial structures of business and trade. Upon his return to South Germany in 1487, he modeled his business practices after those of the Italian city-states. [58]

Fugger defined the LMI of business finance and created an apparently simple and most transparent description of the otherwise complex natural financial behavior of an enterprise. He defined the structures of the income statement, the balance sheet, and the general ledger as we

Figure 3.1 Jakob Fugger (Internet)

29

know them today. As he tracked his own businesses according to the novel financial structures, Fugger became the richest person in the world, and he would still be the richest parson by today's standards.

Fugger utilized his Italian business experience and the discussions with his colleagues in Italy's banking centers in Genoa, Florence, and Venice. He learned primarily from the Italian banker, Luca Pacioli, who, in 1494 AD, published his work on double-entry bookkeeping, including the all-controlling accounting equation, debits, credits, and balances.

Fugger carefully observed the financial dynamics of his businesses and identified logical aspects, which disclosed to him the repetitive natural business behavior consistent with Luca Pacioli's principles. He recorded and tracked his business transactions accordingly.

The Income Statement

The income statement represents revenue and cost for a defined operating period. Depending on management's needs and reporting objectives, the operating period can be a week, a month, a quarter, or a year. Most practical and common is the month selected as the monitoring and reporting period.

One of Fugger's discoveries is that there are two different components of the overall cost experienced by an enterprise. The variable costs (VC) combine the cost of items that are directly caused by the production of the products or services of the enterprise. These costs are also referred to as the direct cost. The variable costs are proportional to the number of units sold, just like the revenue (REV) is proportional to the number of units sold. Consequently, the proportionality between variable costs and revenue is always the same, independent of

```
Figure 3.2   Income Statement per Fugger

Revenue (REV)            = Unit Price * Quantity Sold
Variable Cost (VC)       = Unit VC * Quantity Sold
Gross Profit (GP)        = REV - VC
Gross Profit Margin (GPM) = GP / REV
Fixed Costs (FC)         = Fixed
Net Operating Income (NOI) = GP - FC


NOI is often referred to as EBITDA
EBITDA - Earnings Before Interest, Taxes, Depreciation
         and & Amortization
```

the level of production and revenues achieved. This proportionality is expressed by the Gross Profit (GP) and the Gross Profit Margin (GPM). For a well-managed enterprise, the GPM has to be constant from month to month, as long as the basic enterprise configuration and its financial structure remain unchanged.

The GP and its margin are critically important measures of the financial quality of enterprise operations. For example, if the fixed costs amount to 35% of revenue, the GPM must amount to at least 35% so that the Net Operating Income (NOI) can be positive. Further, the GPM's absolute maximum value would be 100%, which implies that the enterprise would have no variable cost. The highest practical GPMs are in the 75% to 90% range.

Accountants prefer to refer to the GPM as the Contribution Margin (CM). The CM indicates to them how well the operation contributes to covering the fixed costs before further sales begin generating profits each month. The GPM characteristics are illustrated by the GPM line in a diagram where the Net Operating Income (aka EBITDA) is plotted versus Revenue. At zero sales, the losses of an enterprise are marked by the fixed costs. As sales build up, the EBITDA line crosses the breakeven condition and enters the profitable range. The slope of that EBITDA-line is determined by the GPM (aka CM).

Venture capitalists like to refer to the GPM as the earnings Potential (EP). As shown in the chart, the GPM-determined slope of the EBITDA-line is constant across the full range of revenue, including the low revenue range where the EBITDA is still negative. However, considering that GPM is constant across the full range of revenue, the GPM at low-revenue and negative profit condition of startup ventures is indicative of the earnings potential when the revenue begins to rise beyond the breakeven point. The GPM determines at which sales volume the fixed cost can be fully covered so that the enterprise can start generating a positive cash flow.

The fixed costs combine the cost items that are independent of the level of production and the revenues achieved. These costs are also referred to as in-direct costs and as operating expenses. Fixed costs are the monthly expenses resulting from providing the capacity of operations

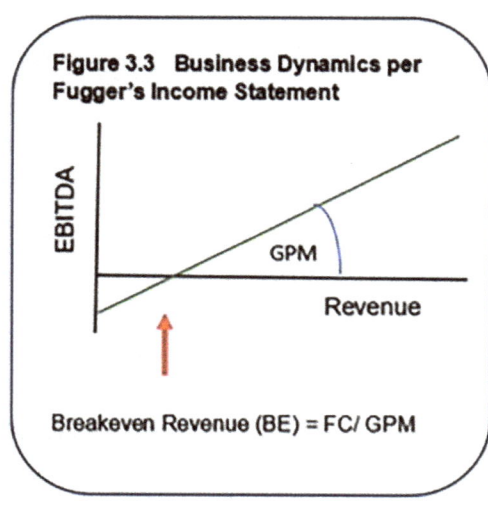

Figure 3.3 Business Dynamics per Fugger's Income Statement

EBITDA

GPM

Revenue

Breakeven Revenue (BE) = FC/ GPM

and production of the products and services. They include, for example, the expenses for office space, insurance, and interest on business loans.

The Logical Mathematical Intelligence associated with the Income Statement reflects an enterprise's natural financial behavior. In the center of that intelligence is the all-important Gross Profit Margin. The main purpose of planning a business and monitoring its performance is to create a strong GPM and monitor it constantly. An enterprise will follow that intelligence, if one knows it or not. For example, when a company operates at a loss, one might conclude additional sales at the amount of that loss would eliminate the loss. That opinion neglects the fact that losses mean that some of the fixed costs are not yet covered and that only the gross profit portion of revenues applies to cover fixed costs. Thus, the incremental sales revenue required to cover the remaining loss is determined by the prevailing GPM.

Incremental Revenue Requirement   =   Loss / GPM

Depending on the GPM value, the incremental revue required to cover any remaining losses may amount to twice the remaining loss.

The Balance Sheet

The second Logical-Mathematical intelligence defined by Jakob Fugger involves the balance sheet. While the income statement specifies revenue and related cost for a certain period, the balance sheet is designed to specify the financial status of an enterprise at a certain point in time. This is done by summarizing credit and debt positions to establish what the owner of the enterprise actually owns. The basic structure of the balance sheet in use today is consistent with Jacob Fugger's thought experiment and his LMI that he established in the 1500s.

The balance sheet is structured according to the accounting equation:

$$Assets \quad = \quad Liability \quad + \quad Equity$$

Figure 3.4	Balance Sheet per Fugger
Assets	Liability & Equity
Current Assets	Current Liabilities
Non-Current Assets	Non-Current Liabilities
	Equity
Total Assets	Total Liabilities & Equity

The assets reflect monetary values, equipment, real estate, and buildings acquired to establish the operational capacity. Current and non-current reflects the time span that would be required to sell the assets for cash.

On the liability side, we find liabilities and equity consistent with the Accounting Equation, which specifies that liabilities and equities belong together, i.e., they belong on the same side of the balance sheet. This makes good sense. Liabilities and equity are of the same nature. Liabilities are owed by the enterprise to external entities while equity is the sum of earnings owed by the enterprise to the owners of the enterprise.

The monetary values of assets and liabilities are typically obtained from bookkeeping records. According to the Accounting Equation, the sum of equities is basically the difference between total assets and total liabilities, which creates the overall balance of all accounts.

The General Ledger

The third LMI concept developed and introduced by Jakob Fugger involves the General Ledger. This is a list of all accounts, which reflect all financial transactions conducted by the enterprise. Fugger's general ledger intelligence introduces the double accounting principle, which specifies that each account has a debit and a credit side, and each transaction must be recorded twice, i.e., where the funds to be transferred come from and where the funds shall be transferred to.

As the ledger reflects essentially all transactions, the ledger must be structured such that it can function as the data source for the enterprise's income statement and its balance sheet.

Cash Flow Statement

Fugger's fourth LMI concept involves the cash flow statement. This was particularly important for Jakob Fugger because of his enormous financial transactions, investments, and donations. The cash flow statement lists cash obligations and cash receipts. The purpose is to ensure that all future cash commitments are consistently covered by readily available cash.

Fugger's Business Operations Prove the Value of his Accounting Practices

Because of his expertise in managing his finances, Fugger became famous for his lucrative business practices, his wealth, and his role in the rise of the Fugger Family. His wealth allowed him to loan money to the Holy Roman Emperor and other royals, which helped him in return to gain power and influence. Fugger's formidable mercantile power was founded on rock-solid security of his metal-backed loan business and his ruthless manipulation of markets.

While pursuing his business practices, Jakob Fugger organized financial reporting structures according to his LMI and achieved a most effective approach to business management. He created modern accounting practices, and, through his consulting efforts, he instigated a rapid spreading of his methodology. His financial success gained him the nickname Fugger "The Rich." He is still reckoned to be the single wealthiest person ever to have lived. Fugger had several thought experiments, one for each of his different business endeavors. He executed his dreams vigorously, but the dreams alone and their vigorous executions by themselves were not the sole reasons for his unprecedented success. He also combined existing knowledge in accounting and created a new LMI that would enable him to manage his businesses so very successfully.

Fugger had a thirst for information about trade and commerce that led him to create a network of couriers whose reports to Augsburg were printed and distributed to clients in the form of a primitive newspaper. Thus, Fugger invented the world's first news service.

Figure 3.5   Image of the *Fuggerei*, Founded by Jakob Fugger in 1516 in Augsburg, Germany. The World's First Social Community.   (Internet)

Fugger had a good heart and created exceptional social programs for the general public. For example, he created in his hometown Augsburg, Germany a low-income housing community, called the *Fuggerei,* which still exists today.

On the other hand, some of his business practices were shrewd. Once he flooded the market with so much metal that the price collapsed, and his competitors were gravely weakened. He used his financial gains to help finance a Portuguese scheme to relocate the pepper and spice trade to Lisbon, a move so successful that it delivered a fatal blow to the commercial stature of Venice.

The financier Fugger raised fresh capital for his bank by exploiting savings accounts, which were first introduced in Augsburg and paid 5% a year, thus contravening the Catholic church's ban on usury. Fugger took his argument directly to Pope Leo X, who had personally benefited from his largesse. The pope was sympathetic, and the ban on usury was conveniently rewritten in 1515, when the process was redefined as "a profit that is acquired without labor, cost or risk." Taking risks with clients' savings had become a legitimate business. It has been said that Fugger paved the way for the modern economy.

Fugger's relationship with the Vatican was based on an extensive network through which he could transfer offertory collections from Germany to Rome (in return for a 3% commission). But his most astonishing and unexpected historical achievement, was unintentionally lighting the fuse that started the Reformation. Fugger teamed up with another Habsburg client, for whom he had bought the archbishopric of Mainz, and the pair began to sell indulgences (a forgiveness of sins, which provided, for a fee, a short cut to heaven), splitting the proceeds with Pope Leo, who used the cash to build St Peter's Basilica. In 1517, Martin Luther was sufficiently outraged by this scheme that he wrote the "95 Theses" that damned Rome, sparking the Protestant Reformation.

Jakob Fugger, with his disciplined logical-mathematical intelligence for assessing and managing his business-related thought experiments, set a formidable example for entrepreneurs in this digital age.

The Natural Causes of Wealth Generation Defined by Adam Smith

Adam Smith (1723–1790 AD) was a philosopher and is considered the founding father of classical economics. He taught logic and moral philosophy at the Glasgow University and published his lectures, of which his 1776 publication, titled *The Wealth of Nations*, is the most notable one. [46] It is the most comprehensive work on fundamental economic theory covering more than 1,000 pages. Smith's theories in economics reflect his observations gained from exploring extensive data and from applying his logical philosophical thinking. His theories are fundamentally sound, have formed the basis of classical economic theory, and affected the works of influential economists such as David Ricardo (1772–1823), Karl Marx (1818–1883), John Maynard Keynes (1883–1946), Milton Friedman (1912–2006), and Paul Samuelson (1915–2009).

The application of Smith's theories and their subsequent solidification by the works of said economists nicely match the Synergetic Five-Step Innovation Process. Adam Smith provided the thought experiment, and subsequent economists provided the logical-mathematical intelligence, which substantiated Smith's thought experiment and provided indisputable proof of his theoretical arguments.

Smith was guided by the fascinating observation that self-interested behavior can lead to the good of the whole. He recognized that a mysterious force is present in the free market that balances the competition among self-interested sellers against the behavior of buyers who seek the best possible deals for themselves. Smith calls this mysterious force the "invisible hand." The latter guides Smith's thinking through complex consequences in the marketplace. When suppliers produce too little of something to meet the market's demand, the price will rise; when sellers charge too high a price, others step in and charge a lower one. Smith's philosophical approach was leading him to the theoretical concept of his invisible hand. Subsequent mathematical treatments unleashed underlying logical-mathematical intelligence, the supply and demand functionality.

In search of means to create wealth, Smith introduced the enormous concept of division of labor, which is accomplished by individuals specializing in doing or producing a particular thing. Specialization improves productivity because it allows workers to develop more skills to do things faster and at greater persistent precision. Smith also recognized that the larger the market, the greater the opportunity for specialization is.

It should be noted that the needs for large markets and specialized labor were indeed recognized and practiced centuries earlier in Central Europe by the Hanseatic League and the business tycoon Jakob Fugger. It took the philosophical mind of Adam Smith to combine emerging dynamics into a consistent set of economic theories.

The Natural Laws Governing Supply and Demand

At the center of his work are Smith's thoughts about free markets and their self-controlling forces. His thought experiment specifies an *invisible hand* that guides the forces of supply and demand in an economy. [46] Smith envisions a free market where the *invisible hand* controls the course of events and allows individuals to look out for themselves while achieving the best outcome for all.

Adding logical-mathematical intelligence to Smith's thought experiment illustrates the functionality behind the *invisible hand*. Linear supply and demand, as shown in Figure 3.6, can be defined mathematically by linear correlations of the form:

$$Price_d \quad = \quad -m_d * q + p_d$$

$$Price_s \quad = \quad m_s * q + p_s$$

With m as the slope of demand and supply lines, where q is the quantity and p marks the lines' intercepts with the vertical price axis.

In a free market economy, Supply and Demand tend to reach their equilibrium, i.e., their crossover, which defines a common quantity Q for demand and supply.

$$Q_E = (p_d - p_s) / (m_s + m_d)$$

With

$$p_s = 0$$

one obtains

$$Q_E = p_d / (m_s + m_d)$$

LMI can provide information on how to guide the market conditions toward lower prices and higher quantities for the benefit of the society as a whole.

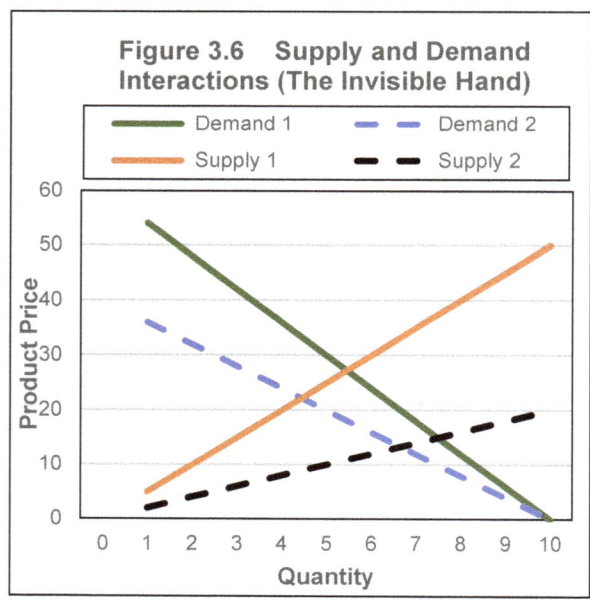

Figure 3.6 Supply and Demand Interactions (The Invisible Hand)

The dotted lines in the chart provide the answer. In order to guide the economy toward higher quantities and lower market prices, the slope of the supply line should be as low as possible, i.e., a small increase in market prices should cause the industry to produce a disproportionately larger quantity. This supply characteristic can be accomplished by high efficiencies in the production process, which typically is accomplished by high labor productivity, which is advertised by Adam Smith in the form of his division of labor. In today's industrial and digital age, high technology content in the production process achieves the same result. Further, the demand line should also feature a

38

low slope and a correspondingly low intercept point p_d, which is typically accomplished by providing strong competition in the marketplace. This is also consistent with Adam Smith's claim defined 250 years ago.

<u>Revenue Potential Associated with a Linear Demand Line</u>

As we recognize that the community of consumers needs low prices and large quantities to raise the societal benefits, the question of how the industry should respond to maximize its benefits arises. The LMI of supply and demand can offer a promising opportunity. The intelligence provides the formula for a linear demand function:

$$Price \qquad = -m * q + b$$

The achievable total revenue associated with a linear demand function is a quadratic correlation that forms a parabola:

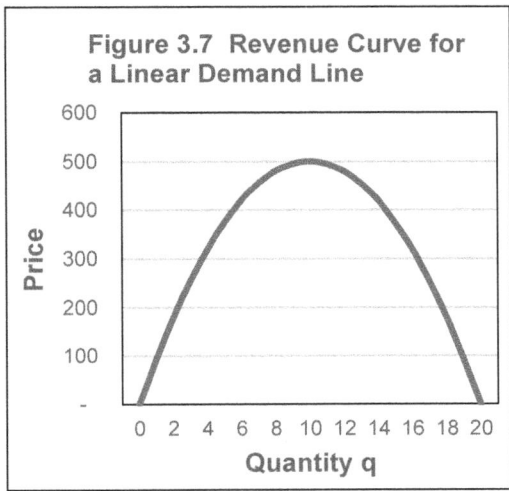

Figure 3.7 Revenue Curve for a Linear Demand Line

$$Revenue \qquad = -m * q^2 + b$$

The revenue curve yields a maximum at the midpoint of the demand line. The marginal revenue, a measure when considering incremental increases in production and sales, is positive and declines until it reaches zero at the revenue curve's maximum. Beyond the midpoint, the marginal revenue is negative. Individual companies need to assess their marginal revenue and determine their operating conditions that deliver the maximum revenue for them.

<u>The Natural Laws Governing the Market Structures</u>

The demand for an enterprise's products and services as described above is affected by the market structures that the enterprises face.

The market's different configurations can be viewed as a continuous spectrum of structures that smoothly interface and transition from one to the other. At the extreme ends of that spectrum, we have the perfect competition and the monopoly.

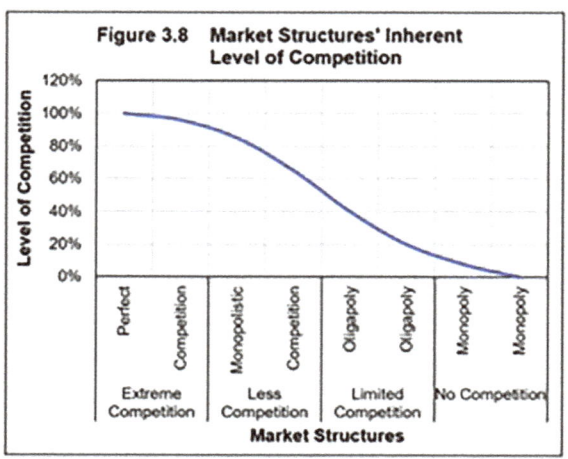

Figure 3.8   Market Structures' Inherent Level of Competition

At perfect competition, the market features a large number of organizations that are in competition with each other. The market entry for new organizations is easy. Moving away from the extreme perfect competition, when the perfect competitive market tends to adopt monopolistic characteristics, the market experiences less competition.

When the market is dominated by a small number of large corporations, we have an oligopoly condition, which offers limited competition and great difficulty for new organizations to enter the market. A monopoly condition is reached when the market is dominated by a few or even a single large corporation. All forms of market structures can be comprehended when the characteristics of the two extreme market structures, i.e., perfect competition and monopoly, are discussed.

Perfect Competition

Perfect competition is achieved when the market is formed by a large number of suppliers who are in strong competition with each other. No single company dominates the market. All suppliers are small businesses. Each supplier is a small fish in a big pond. As a single small supplier increases his production and pushes his units into the market, his additional units have no impact on the market as a whole. Consequently, the market price will appear to him as a constant market-given price, which is reflected in the chart by a horizontal line.

An organization wants to explore optimum production, and there is a logical natural guideline. When organizations increase their production

40

unit by unit, the incremental units typically yield increased cost, i.e., the marginal cost rises. Each incremental unit produced generates an incremental revenue. Under perfect competition conditions, the marginal revenue is equivalent to the price provided by the horizontal demand function. Thus, for perfect competition, the maximum profitable production is attained at a quantity production, where the MC-curve intercepts the horizontal price-line.

Marginal Revenue (MR)   =   Marginal Cost (MC)   =   Price

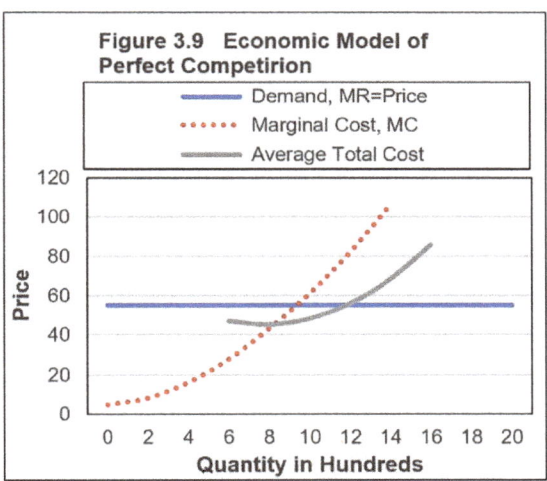

Figure 3.9 Economic Model of Perfect Competirion

At the MR vs MC crossover, the total cost must be lower than the MR price line to ensure profitability of the organization. This is exactly where the perfect competition situation causes severe problems. In perfect competition, the market price is constantly undercut by competitors who want to gain market share. As a result, the market price tends to decrease. In addition, the average total cost has a tendency to rise so that the profitability is being squeezed from both ends and is constantly declining.

The business entry into a market that features perfect competition is relatively easy. There are no significant barriers. Consequently, the large number of suppliers causes the lowest acceptable price and satisfies a large number of consumers. From the consumers' perspective and that of the entire society, perfect competition is the preferred market structure.

Due to the profitability of perfect competition, organizations are constantly under severe pressure, with owners constantly being challenged for improvements. Under perfect competition, business owners must constantly look for ways to reduce cost and to engage in innovations for their processes and for completely new products. Their most desired objective is to invent new products that would take their organizations out of the realm of perfect competition and offer them the benefits associated with monopolies.

Monopoly

While perfect competition entails a large number of suppliers who each face an apparent horizontal demand line, the monopoly entails very few suppliers (or only one supplier) who dominate the market and who face the real demand line. If a monopolist changes his output by a certain percentage, the incremental quantity represents an equivalent change of the entire market, and the price responds accordingly. This market response results in a marginal revenue line that runs below the market demand line and reaches the horizontal quantity axis at the demand line's midpoint. This unique characteristic is a result of the total revenue curve, which reaches its maximum at the demand line's midpoint. These features result in an enormous economic benefit for monopolies over perfect competition.

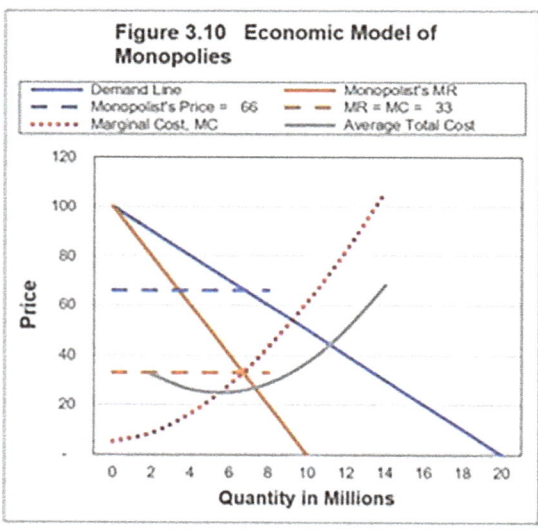

Figure 3.10 Economic Model of Monopolies

- Demand Line
- Monopolist's Price = 66
- Marginal Cost, MC
- Monopolist's MR
- MR = MC = 33
- Average Total Cost

The monopolist conducts an assessment of his production costs and determines for his facility the marginal cost curve. Using the production criteria

$$MR = MC$$

he determines the quantity that he wants to produce. However, in contrast to perfect competition, the point $MR = MC$ does not specify the price that he can charge. The price is specified by the demand line so that the actual price that the market is willing to pay is much higher than that given by the MR vs MC crossover condition. The monopolist's superior benefit is marked in the chart by the brown and blue dotted lines. This is the extraordinary benefit offered by a monopoly.

42

A monopolist is typically accused of setting excessively high prices to exploit consumers. One must understand, though, that the economic model discussed above is a natural model built on LMI. It is not a man-made configuration. The monopolist's economic model is based on the fundamental functionalities of the monopoly, which function as described whether we know those natural laws or not. The question is not the model, but under which conditions is the model allowed.

Since the monopoly is the market structure that is desired by the business community over perfect competition, business owners would try to achieve monopolies any way they can. This trend is one of the instabilities existing in economics. In the 1940s, monopolies represented 80% of the US economy. Rules have been established that disallow man-made monopolies. It is not allowed, for example, for like businesses to join forces in order to create monopolies. Under modern business laws, only natural monopolies are allowed that emerge when ingenuity has created new products and/or new processes, which support superior competitive advantage in the marketplace. And that is exactly the reason why the monopoly model is still allowed. It is a carrot for the many perfect competitors. It is a motivator for innovation and creativity.

The concurrent existence of perfect competition and monopoly creates an ideal power play in the economy. The many perfect-competition organizations form an economic basis where the wealth of the society is generated. Occasionally, a perfect competitor launches an innovative new product and enjoys the benefits of the natural monopoly that he created. With time, competitors find a way to create competition, and the monopoly transforms back into perfect competition.

According to Adam Smith, the forces of wealth building are the division of labor, by which individuals specialize in producing particular things, and the invisible hand, which allocates resources effectively. These fundamental features of free capitalist economies have not changed during the last 250 years. In fact, the transitions into successively deeper developments into industrialization and into the digital age have confirmed the validity of Smith's theoretical findings. Practical experience and the emergence of geometrical displays and algebraic formulations have delivered the strongest proof of Smith's theoretical deliberations.

Enterprise Performance Strategy

In the center of Fugger's transparent income statement are three monetary components that form the basis of an enterprise's financial performance. Earned revenues are applied to variable and fixed cost, and the remaining balance makes up the operating income (EBITDA). A two-dimensional chart that illustrates the net operating income (EBITDA) as a function of the full ranges of variable and fixed costs is shown in Figure 3.11.

In the chart, costs and profit are represented in percent of revenue. To obtain a meaningful business performance diagram, the variable costs are represented indirectly in form of Gross Profit Margin:

Gross Profit Margin　(%)　　=　　Revenue (%) - Variable Cost (%)

Variable Cost (%)　　=　　Revenue (%)　- Gross Profit Margin (%)

Accordingly, when the Variable Cost (VC) amounts to 100% of revenue, the Gross Profit Margin (GPM) is equivalent to zero, and vice versa. Consequently, when VC and GPM attain the same percent of revenue, the enterprise's EBITDA is equivalent to zero i.e., the enterprise operates at breakeven.

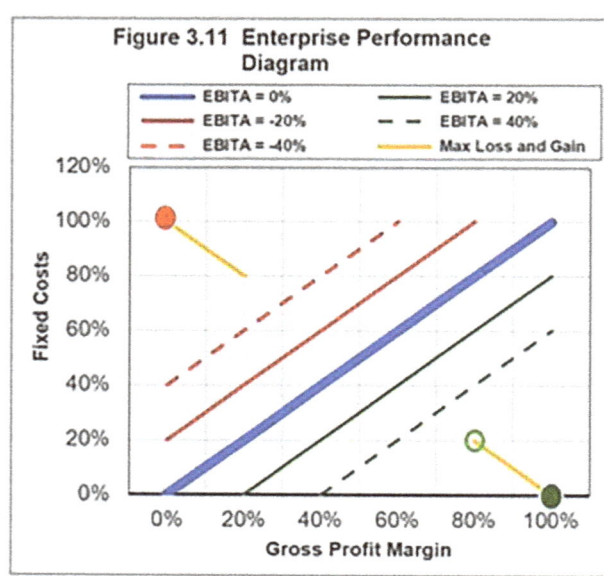

Figure 3.11 Enterprise Performance Diagram

In the diagram, the solid blue line marks the FC and GPM combinations that result in EBITDA values equal to zero (Breakeven Condition). Thus, the solid blue line divides the entire chart into two performance regions, i.e., negative and positive EBITDAs. The solid and dotted red lines mark EBITDA equivalences of negative 20% and negative 40%

of revenue, respectively. The solid and dotted green lines mark EBITDA equivalences of positive 20% and 40% of revenue, respectively.

The worst operating condition included in the chart is marked by the red dot, where the variable cost and the fixed cost amount to 100% of sales (GPM = 0%), and the EBITDA would amount to -100% of revenue. The absolute best operating condition, although very unlikely, is marked by the green dot, where both variable and fixed cost amount to zero (GPM = 100%) and EBITDA would amount to 100% of revenue.

A more practically feasible yet extremely high enterprise performance is marked by the green circle, where variable cost is 20% (GPM = 80%) and the fixed cost is also maintained at 20% of revenue. In that case, the EBITDA would amount to 60% (80% - 20%).

Management and leadership teams can enter their enterprise's performance and those of their competitors in the chart and make a visual comparison of their financial competitive position in the marketplace. Further, the chart can assist in defining a financial strategy for approaching the green circle as close as possible, while drastically improving their competitive position in the marketplace.

The Enterprise Performance Diagram represents critically important logical-mathematical knowledge established with the LMI introduced by Jakob Fugger in the 1500s.

Interplay between Market Structures

The course of the events in economics over the past 250 years has

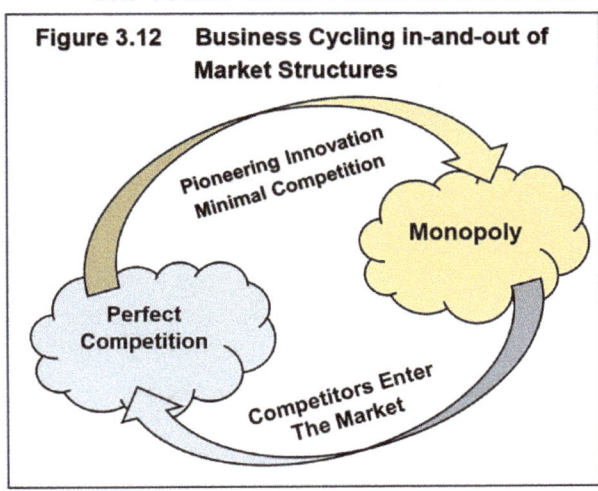

Figure 3.12   Business Cycling in-and-out of Market Structures

Pioneering Innovation
Minimal Competition

Monopoly

Perfect Competition

Competitors Enter The Market

illustrated that the individual's motivation for accomplishing more, for finding new ways to satisfy the unlimited human desires, is a dominant factor driving the economy in the direction defined by Adam Smith. and consequential motivation.

As the organizations operating in the field of perfect competition yield constantly diminishing profits caused by market pressures toward lower prices and by supplier pressures toward higher cost, business leaders are forced to constantly search for new products and processes.

With new intellectual properties (IP), a company can create innovative products and services and can temporarily enjoy the financial benefits inherently provided by monopolistic market structures. The monopolistic status can remain until competitors develop comparable products and enter the market. Thus, the task at hand is not just to develop new IP for new products and services but to unleash monopolistic benefits. Furthermore, leadership also has to ensure that the monopolistic market position can be sustained for an extended period.

Exercising the Synergetic Five-Step Innovation Process, an enterprise can create firm and sustainable monopolistic conditions and extend the economic privileges associated with monopolies. Existing LMI must be fully exploited to assess opportunities and risks and solidify a straight path to non-wavering success. If the required LMI does not exist yet, it needs to be developed. In any event, it is critically important to attain the LMI to fully understand success and failure modes in the market and to solidify the defensibility of the enterprise's innovative IP that underlies the new products and services.

The Wealth Generation Intelligence

In his 1776 publication, *The Wealth of Nations*, Adam Smith emphasizes several thoughts on productivity as a source of wealth. [46] Smith's thought experiment caused economist Paul Samuelson to derive applicable logical-mathematical intelligence that supports Smith's claim. [41]

Starting with LMI, which states that the market price of a product is proportional to the production labor cost, gives us this correlation:

$$Price \sim (Hours * Wage) / Production$$

With

$$Productivity = Production / Hours$$

One obtains the price being proportional to hourly wages per hourly production. The latter is equivalent to labor productivity.

Price   ~   Wages / Productivity

With the mathematical transformation of price, wage, and productivity to annual rates of change, one obtains a correlation that connects the annual rate for price inflation with the annual rates of wage and productivity growth.

Price Inflation   =   Wage Growth - Productivity Growth

The LMI applicable to Adam Smith's thought experiment illustrates that the annual productivity growth creates sustainable growth of a nation's wealth. For example, the green bars in the chart of Figure 3.13 reflect a healthy economy. If an annual wage growth of 7% is counter-balanced by an annual growth in productivity of 2%, the annual price inflation rate will tend to stabilize in the range of 5%. At the end of the year, the labor force's remaining gain can amount to 2%, which is directly equivalent to the economy's growth in productivity.

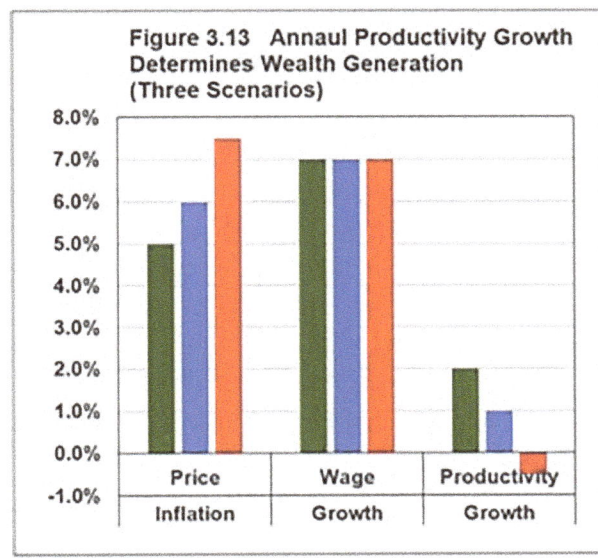

Figure 3.13 Annaul Productivity Growth Determines Wealth Generation (Three Scenarios)

Policymakers often speak of wages' inflationary pressure. They are correct, but that is only half the truth. The applicable LMI reveals that annual productivity growth is the most powerful means for neutralizing inflationary pressure caused by wages.

The three components of the wealth generation formula represent averages of a society's overall economic conditions. Consequently, an economy's total wage volume must be seen as the product of the economy's average wage per capita and the economy's pertaining total employment.

Wage Volume = Average Wage per Capita * Employment

Thus, wage growth in the wealth generation formula is the result of changes in wages and employment. This confirms the well-known fact that increases in wage rates and the reductions in unemployment both tend to drive price inflation upward. We constantly observe pressure to raise wages and reduce unemployment. To maintain price inflation at a level that is lower than the economy's growth in total wages, the economy's productivity must be maintained at a positive rate.

Productivity growth performs a balancing act and diminishes the adverse impact of wage growth. It is important that the annual price inflation is less than the annually occurring average rise of wages. This difference between wage growth and price inflation is created by productivity growth. Consequently, the growth of a nation's wealth is simply created by the nation's annual productivity growth. Conditions in which a nation's productivity growth rate is nil or even negative (red bars in Figure 3.13) are highly undesirable. This observation obtained through the use of LMI is consistent with Adam Smith, who declared, *"Wealth is created through productive labour."* [46]

The wealth generation formula reveals an important aspect of an economy's dynamics. For example, when a society's productivity growth is low or even nil, the entire society will experience high inflation and will request wage increases to compensate for the inflation. However, the economically unjustified wage growth will contribute to further declines in productivity and will make things worse for all. This sequence of events creates a self-feeding instability of the economic system from which it is difficult to recover. Such instability must be avoided in the first place by focusing on productivity growth.

A society's goal must be to focus on productivity and a healthy annual productivity growth. A raise in wages must be earned by creating productivity growth. Thus, the sequence of events must be different than the one that is often practiced. To create and maintain a stable and healthy economic system, productivity must be raised first to justify growth in

48

wages. This concept adds critically important social responsibility to every business management team's traditional set of corporate responsibilities. Fortunately, this social responsibility is well in line with traditional corporate responsibilities and simply puts extra emphasis on the team's responsibility for creating and maintaining a healthy annual growth in productivity.

The Synergetic Five-Step Innovation Process with logical-mathematical intelligence at its center is meant to assist management teams in accomplishing the all-important objective of "Productivity First."

Unique Barter and Money Systems' Economies

The phenomena surrounding the money concept are part of the complex social science of economics. The human players in the game can be viewed as the manipulators of a complex set of natural laws. Consequently, just like in sports, the human players have to understand not just the man-made rules, but also the underlying natural laws at play. The latter just cannot be altered or violated.

This author tends to view economic phenomena through the eyes of a natural scientist who thinks in terms of physics, engineering, and mathematics. He judges from first principles and not by analogy. With that in mind, he wants to look at some of the naturally bound facts underlying money related phenomena.

The creation of monetary systems is due to a natural law combined with the human brain's logic for efficiency. A primitive alternative to a money economy is a barter economy. In fact, a barter economy may be seen as a temporary alternative when money has been withdrawn from the economy or when the currency has become useless because of political uproars. A barter economy is so inefficient that the human participants in that economy tend to start practicing a money-type economy. The natural trend toward a money-based economy can be illustrated by LMI.

Consider a barter system with 10 participants, and each participant has multiple units of one unique product. If a participant wants to exchange several units of his product against several units of another product owned by another participant, he needs to check all other trade options within the system to ensure that he gets a fair deal. Consequently, all desired exchanges between all 10 participants results in 10 times 9 communications considering each connection in both directions. The division by 2 determines the trade options.

$$\text{Trade Options} = (10 * 9) / 2 = 45$$

Using N for the number of participants in the system, one obtains the general formula

$$\text{Trade Options} = N * (N-1) / 2$$

When N is a large number, the "1" can be ignored so that the general formula becomes a simple exponential function for a barter-based trading system.

$$\text{Trade Options} = N^2 / 2$$

Figure 3.14 Trade Options in Barter and Money-Based Economies

Barter System	Money System
Direct Person-to-Person Exchange	Single Central Market
Multiple Contacts	Single Contact
Trade Options = $N^2 / 2$	Trade Options = N

For a money system, let's assume that one of the system's products is used as a commonly accepted reference of value. The reference unit becomes a central object that everyone accepts in return for a sale so that it can subsequently be used to purchase other products available in the system. The single commonly accepted product eliminates the ineffective cross-reference with all available products in the system to determine the best deal. Thus, each trade requires just one reference point,

50

i.e., each product's value is determined in reference to the single commonly accepted product. In a 10-product system, the trade options for each product reduces from 9 to 1 because of the commonly accepted single value product, which we call money. Thus, the barter formula reduces to a single central market system's formula with a single reference product, which is illustrated in the chart.

The money system discussed above, where sellers sell in a central market and buyers purchase their goods in that central market, has become known as the hub-and-spoke system. Because of its striking economic superiority over a non-hub system, the hub-and-spoke system is implemented wherever it is practically feasible to do so. In a broader sense, the hub ensures that all spoke organizations are working synergistically together toward a common goal, minimizing duplication of efforts and promoting alignment of strategies.

The hub-and-spoke system has become the preferred network design for industrial organizations and their supply as well as their distribution networks. The hub system is also applied in physical system designs. Plumbers include manifolds when they are dealing with large networks of pipe systems, and electronic designers include central hubs for electronic data collection and distribution.

Interestingly, the human brain makes reference to the advantages of the hub system and creates a logical mindset guiding humans to engage in hub and spoke behavior even though the conscious mind may not necessarily know of the numeric benefit of the hub versus the non-hub system. For example, it has been reported that, some 10,000 years ago, tribes in Africa, used rare seashells as commonly accepted "money" to smoothen their trade transactions.

Natural Laws of Money Systems' Inherent Money Multiplication Phenomena

The previous section illustrated that at the heart of economics is money. Money is earned, saved, and spent. Important aspects of money are the consequences of changes in money supply. Money supply affects the fundamental objectives in managing the economy and interacts with a nation's economic growth, its price inflation, and unemployment status.

Money management is a very sensitive balancing act. On one hand, the economy needs access to a readily available supply of money; on the other hand, over-supply causes inflation and unemployment and has caused catastrophic outcomes like the Great Depression.

Money initially played a simple, practical function in the economy. Although its classical function as a means for making payments and for storing value is still in effect, the concept of money has adopted significant intangible features as the economy has grown and become more sophisticated and complex.

The very first intangible deed accomplished by money happens at the very inception of money. Its first deed is to convert the cumbersome moneyless barter economy to the more efficient money-based economy. This transition is natural and is driven by humans' logical perception of an easier solution when barter-type conditions are unacceptably difficult and cumbersome. The natural logical perception of solutions is typically directed by how it creates personal benefits. These personal desires are beneficial for the society as long as they are thriving, for example, more effective ways of doing things with the objective to raise personal gain, as was pointed out so well by Adam Smith. But when it comes to money, personal interests do not necessarily benefit the society as a whole.

Money can serve its beneficial purposes as long as the money itself is scarce. When people used natural resources like shells, gold, and silver for money, there was a natural limitation of money supply. But when goldsmiths who were storing gold for their clients and issued paper certificates for those gold holdings and when people started using their certificates for purchasing goods and services, the gold certificates began to function as money. The goldsmiths charged a small percentage for holding gold and developed the thought that their fees for gold holdings were actually fees for the certificates they issued. From that thought it was only a minor step for the goldsmiths to start selling loan certificates based on the gold deposits they were holding. The next thing the goldsmiths observed was that since people were using gold certificates for payments, few clients withdrew gold deposits so that the amount of physical gold deposits were quite stable. As a result, the goldsmiths felt that it was safe to issue loan certificates equivalent to more than the value of their gold holdings in their vault. With that step, the money-multiplying concept of the modern financial systems was born.

The financial incentive for issuing loans in terms of gold certificates became very obvious to the goldsmiths at the time. So much so, that their profession changed from goldsmithing to banking.

If the gold deposits are equivalent to G, and if the issued loans are equivalent to L, the Money Multiplier (MM) amounts to:

$$MM \quad = \quad L/G \quad = \quad (L/L)/(G/L)$$

The expression (G/L) marks the ratio of the amount of gold deposits (G) used to back-up the loan amount (L) issued in the form of gold certificates. In modern banking, this ratio is called reserve rate (RR). Thus, we obtain:

$$MM \quad = \quad 1/RR$$

This is equivalent to the dangerous One-Over-X function. If we assume that the banks wanted to limit their issued loans to ten-times of their gold deposits, they would practice a reserve rate of 10%. If they were more aggressive and reduced their reserve rate to 5%, their outstanding loans could be doubled. If every bank used the same practice, the entire economy's money supply would double.

For the goldsmiths who turned into bankers, this was a golden opportunity. If they limited their gold certificates to 100% of the gold deposits, they would earn their 3% fee once on all deposits. If they extended their loans to ten-times of their deposits or even 20 times, they would earn 30% or even 60%, respectively, on their gold deposits. Thus, the earnings of the goldsmith bankers depended on the money multiplication that they practiced.

In the literature, this money multiplication factor is discussed using several value assumptions for the reserve rate, such as, for example, 10% or 5% to illustrate the money multiplication. But the full Logical-Mathematical Intelligence offered by the One-Over-X type money multiplication formula is not discussed. For instance, if in critical economic situations the reserve rate RR is allowed to slip to extremely low values, the money multiplier MM will rise disproportionately and drive the economy into uncontrollable conditions.

In our modern societies with complex financial systems, nations' central banks set their reserve rates and control them carefully at approximately 8%. However, complex economies have other features that have indirect impacts on the money supply and, thus, can indirectly impact the effective money supply.

The banks' debt system, which is directly coupled to the monetary system and based on cash deposits in the banks' vaults, is not the only system available for affecting a nation's money supply. Lending money is coupled with debt creation and uses certain assets as guarantees.

This leads us to an extension of the money multiplication consideration. Using the money-multiplier concept, the money supply can be defined as:

$$\text{Money Supply} \quad = \quad A * 1/RR \quad = \quad A/RR$$

A shall represent the sum of all assets that qualify as deposits and, thus, form the basis for the actual money supply in an economy. If policy-makers create additional sources for debt that can qualify as deposits, the supply can expand in a multiplied fashion.

$$\text{Money Supply Growth} \quad = \quad (A_2 - A_1)/RR$$

Consequently, the stability of the money supply requires disciplined control of the banks' reserve rate and debt creation mechanisms.

A nation's money supply mechanisms must consider three significant constraints. The first constraint involves the banking system's applied reserve rate. Second, the scarcity requirement of money sets a limit on the loan and debt creation process. And the third constraint is that money needs to be earned and shall not be made available to the public without limits. In contrast to the reserve rate and the debt creation mechanisms, which are explicit factors that can be defined in mathematical terms and can be controlled through explicit regulations, the scarcity concept and other indirect impacts on the money supply must be constrained by an appeal to ethical behavior of policymakers and the banking industry administrators.

In addition, as technology changes and opens up new avenues in banking procedures and money management concepts, economic policymakers must apply available LMI to explore new emerging procedures and identify economic instabilities before they happen and before they can cause major catastrophes.

Chapter 4
Logical-Mathematical Intelligence in Natural Sciences

Introduction

This chapter is about the creation of outstanding scientists who were interested in the nature of fundamental geometries (Pythagoras, Euclid), the physical behavior of objects in their environments (Archimedes), fundamental mechanics (Newton), and light and the universe (Einstein). They used their exceptional strengths in logical-mathematical intelligence and established outstanding logical-mathematical knowledge that changed the world and laid the foundation for the industrial age.

During the golden age of philosophy in ancient Greece, Pythagoras built the bridge between geometry and algebra. He was driven by the thought experiment that arithmetic rather than geometry can provide the truth. Archimedes had the advantage of all the knowledge assembled at the largest library of his time, the Central Library of Alexandria. He designed several useful devices such as the water pump and is considered the founder of fluid mechanics.

2,000 years later, during the 17th and 18th centuries, where modern scientific knowledge was at its early stage, Newton's thinking was influenced by philosophers of his time. He studied Latin and Greek philosophy, and later in life he was able to follow his interests in natural science and self-educated himself in mathematics and physics to the extent that he invented calculus and published his work entitled, *Philosophae Naturalis Principia Mathematica.* [34]

200 years later, Albert Einstein started his career by enrolling at the Polytechnic School in Zurich to study physics. Like Newton, most of what he learned was self-taught by reading scientific papers on cutting edge theoretical physics. After graduating and after several years of odd jobs, Einstein got a job at the Zurich patent office as a technical expert where

he found the time to pursue his own research interests in theoretical physics.

It is interesting to note that Pythagoras, Archimedes, Newton, and Einstein developed very simple but unique thought experiments that set the direction for their research. Applying LMI led them to the knowledge that they were looking for. The LMI that they needed did not exist at the time, so they developed it on their own.

The Inventions of Pythagoras

Although working under strict secrecy within a fenced-in enclave, Pythagoras' algebraic discoveries became fundamental elements of progress during the antique age. The Pythagorean idea that arithmetic rather than geometry can provide the truth was adopted by philosophers in ancient Greece and has been carried over into today's high- tech World. Pythagorean Logic is so fundamentally correct and powerful that it is applied in today's most complex research work. And for the entrepreneur, Pythagorean inventions can provide decisive logical-mathematical intelligence for the assessment of pioneering thought experiments.

Figure 4.1 Pythagoras (Internet)

Pythagoras' inventions have become fundamental tools of physicists and engineers. The knowledge involved has reached such a high level of acceptance that its discussion seems to be a superfluous undertaking. However, the underlying logic that Pythagoras employed to uncover the now so obvious laws is striking and may assist young entrepreneurs to fully understand that simplicity can lead to simple yet very powerful and far-reaching innovations. Most of today's powerful inventions with broad commercial value can indeed be reduced to an underlying simplicity. The value of an innovation does not lie in its complexity but in its simplicity. This observation shall be illustrated by referring to Pythagoras' discoveries.

In his search for mathematical correlations that can define geometry,

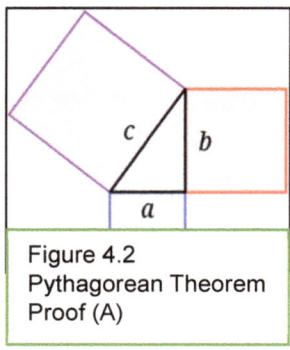

Figure 4.2
Pythagorean Theorem
Proof (A)

Pythagoras and his scholars studied squares and rectangular triangles and defined an overwhelmingly simple algebraic correlation. The correlation is known by many but the logic behind its derivation may not be known so well, although

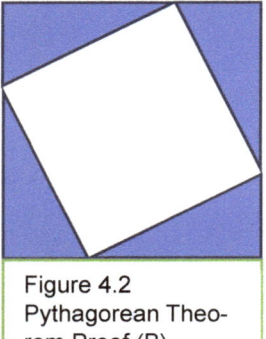

Figure 4.2
Pythagorean Theo-
rem Proof (B)

the geometric and algebraic logic applied by Pythagoras and his scholars can teach us even in the digital age to proceed if we are facing an issue where the logical-mathematical knowledge does not exist yet.

The diagram on the left illustrates the Pythagorean theorem:

$$c^2 \;=\; a^2 \;+\; b^2$$

The diagram on the right illustrates one of the many logical derivatives of the Pythagorean theorem. Four right-angle triangles with the sides a, b, and c are arranged to form an outer square with the area equivalent to $(a + b)^2$ and an inner square with the area equivalent to c^2. The difference between the two squares is defined by the combined areas of the four triangles, which amount to 4 times ½ab.

$$c^2 \qquad = \qquad (a + b)^2 - 2*a*b \qquad = \; a^2 + 2*a*b - 2*a*b + b^2$$

This proves the Pythagorean theorem:

$$c^2 \qquad = \qquad a^2 + b^2$$

The diagram on the left can be used to deliver the same proof. The outer perimeter defines a square with the surface area equivalent to the product (2a +b) times (a + 2b), which must be equal to the some of the internal squares, rectangles, and rectangle triangles.

$$(2a + b) * (a + 2*b) \qquad = \qquad 2*a^2 + 5*a*b + 2*b^2$$
$$= \qquad c^2 + a^2 + b^2 + 3*a*b + 2*a*b$$
$$c^2 \qquad = \qquad a^2 + b^2$$

The Quadratic Expansion

The integral geometric and algebraic proof includes the expression

$$(a + b)^2 \qquad = \quad a^2 + 2*a*b + b^2$$

where the term b^2 is considered the quadratic expansion. If an algebraic analysis of a system of any kind ends in a correlation like

$$a^2 + 2*a*b \qquad = \quad x$$

where the quadratic expansion " b^2 " is missing, the correlation cannot be solved for " a ". Adding the quadratic expansion " b^2 " on both sides of the equation, one obtains

$$a^2 + 2*a*b + b^2 \qquad = \quad x + b^2$$

and now one can solve the correlation for a

$$a \qquad = \quad \sqrt{x + b^2} \ - \ b$$

This process offers solutions for many complex exponential correlations. It is a simple approach to be included in an entrepreneur's portfolio for LMI. A study of the logic behind fundamental derivations enhances logical approaches to the assessment of a thought experiment and leads to independent critical thinking. The astounding observation is that this very effective and often practiced technique in today's digital age relates back 2,500 years to the ancient work of Pythagoras.

The Inventions of Archimedes

Because of Archimedes' Principle, naval architects are able to define the Metacentric Height GM as a ship's stabilizing factor (see the sketch below). The center of gravity (G) and the center of buoyancy (B) create a unique system that behaves like a swinging pendulum. A ship heeling over to one of its sides experiences a shifting of its center of buoyance to that side in such a way that the metacenter (M) is

Figure 4.3 Archimedes (Internet)

located above the ship's center of gravity so that the ship experiences a stabilizing momentum. The length of the section GM, the metacentric height, determines the ships rolling frequency. A large GM creates a strong upright momentum and causes hard narrow swings to each side. A short GM causes less upright momentum and causes wide and soft swings. The naval architect must design the overall ship's structure such that M is located above G with and without commercial loads.

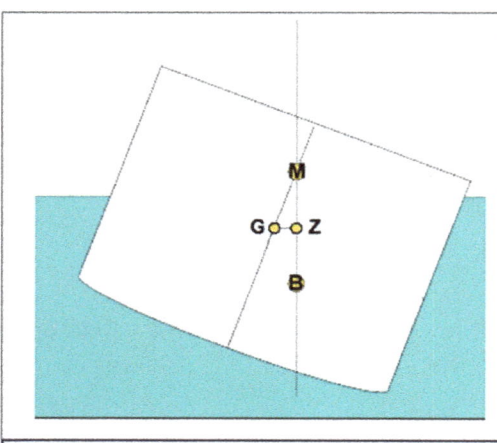

Figure 4.4 The Metacentric Height GM Determines a Vessel's Stability

The GM length is determined by the position of a ship's center of gravity G and by the ship's width, its beam. For ships with heavy steel hulls and with heavy engines in their hulls, the naval architect can design the metacentric height as desired. Ancient wooden sailing vessels had insufficient stability when unloaded. Thus, ships were loaded with heavy stones to enhance stability when they did not have heavy merchant cargo in the hull. The harbor of San Diego in Southern California still features a "ballast point," where early wooden ships could load and/or unload ballast as needed.

Modern sailboats are equipped with a deep reaching keel and a heavy weight at its lower end. The weighted keel creates a low center of gravity and, thus, provides strong stability for the hull. Large commercial steel vessels typically have a low center of gravity and a large beam; both create the metacentric height they need to allow high structures for passenger vessels and high container loads for cargo vessels. Passenger ships are preferably designed with a small metacentric height to achieve a comfortable journey for the passengers. Cargo vessels are designed for larger metacentric heights to provide strong stability and to allow high-reaching cargo loads.

Archimedes' Principle and the metacenter concept provide the LMI to assess the practicality of a naval architect's thought experiment for innovative ship designs. The astounding observation is that this very effectively practiced technique in today's digital age relates back 2,300 years to the ancient work of Archimedes at the Central Library of Alexandria.

Isaac Newton's Search for Logical-Mathematical Intelligence of Gravity Forces

Isaac Newton (1642–1727 AD) was an English physicist and mathematician, who is known for his studies of classical mechanics published in *Philosophae Naturalis Principia Mathematica*. [34]. Just as an example, Newton established the well-known correlation for velocity as the ratio of distance and time.

$$\text{Velocity} \quad = \quad \text{Distance} / \text{Time}$$

His major inventions involve light, calculus, and the correlation that defines gravitational forces. His publication, *Classical Mechanics*, became a manifesto of his philosophical thinking and his practical inventions. [34, 49]

Newton studied the works of Galileo and Johannes Kepler. Galileo had described earthly motion while Kepler had described planetary motion. Newton recognized that a common theory explaining motion was still missing and developed an interest in solving the puzzle.

Figure 4.5. Isaac Newton (1642 – 1727)   (Internet)

During Newton's time, Europe had long experienced the end of the dark Middle Ages and had transitioned into the Renaissance, where the ideas of the Greek philosophers like Pythagoras and Aristotle were no longer sacrosanct. Newton studied Aristotle's thoughts about moving and resting bodies, which had been challenged by Nicholas Copernicus, Galileo, and Kepler. Thinking about motion and the force that creates motion guided him to explore the force that has become known as gravitational force. Using his expertise in calculus and algebra, Newton envisioned a force that would make the moon orbit the Earth and the Earth's planetary motion around the sun. He exercised an interesting thought experiment. He wondered why the moon would not follow a straight path but, instead, circle around the Earth. He imagined a rope that would tie the moon to the Earth and force the moon onto a circular motion around the Earth. Then he tried to determine the force that the rope would have to endure to accomplish the task. Over a period of several years, with the rope-model in mind, he developed the mathematical basis for the theory that eventually led him to the familiar equation for the gravitational force

$$F \quad = \quad G * (M_1 * M_2) / R^2$$

F is the gravitational force dependent on the two involved bodies' masses, i.e., M_1 and M_2, and R is the center-to-center distances between the two bodies. The factor G is the gravitational constant, which was discovered later by Henry Cavendish in 1798. A very important component in Newton's correlation for gravity is his finding that the force's variation with distance is governed by a square law.

The inverse square law determines that if the distance between two bodies is doubled, the gravitational forces would be reduced by the factor 4. Conversely, if the distance between two bodies were to be reduced by half, the effective gravitational forces would be doubled.

Newton's theory of gravity defines that both the celestial motion of the planets and the earthly motion of cannonballs were essentially the same kind of motion except on different scales. Today, Newton's gravity force is known as one of the fundamental natural laws governing motion on Earth and throughout the entire universe. Military ballistics, the aerospace industry, and space exploration are all unthinkable without Newton's gravity correlation. Isaac Newton was the first to use physics to explain astronomical phenomena, which makes him the first astrophysicist.

In order to show that his theory was valid, he had to provide a mathematical basis. Using geometry and algebra, he soon realized that the knowledge of math existing at his time was not sufficient. He needed a way of expressing the rate of change. As a result, Newton invented calculus to evaluate his theory of gravity-affected motion. This discovery confirms the genius mind of Isaac Newton.

There is a side notion to Isaac Newton's work and his life. This genius experienced a disappointing disrespect and neglect by his colleagues at the time. English physicist Robert Hooke (1635–1703 AD), who researched a remarkable variety of fields and discovered the law of elasticity, independently of Newton's gravity related work, came up with similar conclusions about the gravitational forces between two bodies, though he did not explicitly specify that the variation with distance was governed by an inverse square law. This circumstance resulted in discussions of which work is correct and who was first.

There was also a mathematician in Germany, Gottfried Leibniz, who, independently of Isaac Newton's work developed a theory and notation of calculus. A dispute emerged over who had actually invented calculus as a separate branch of mathematics. Although the date of publications indicate that Leibniz was the first, the actual timing of the work performed clearly shows that Newton was indeed the first. The two scientific disputes weighed heavy on Newton's mind and his independent-thinking spirit. He became a bit paranoid about ideas being stolen, which affected his creativity.

The English science community was not prepared to accept his theoretical work and his formula of gravity. He was often confronted with the request to prove his theory and his analytical correlation. Any proof had, of course, to be done by way of a physical experiment. Newton interpreted these repeated requests as a challenge to prove the weight of the Earth. In response, he developed the notion that this is an impossible task and became accustomed to the phrase, "If someone would give me a scale that is large enough to put the Earth on it, I would not know how to do that."

Unfortunately, his phrase became his paradigm, preventing him for the rest of his life from envisioning an experimental proof. A few years after his death, however, a young physicist did come up with an extraordinary experimental arrangement that confirmed Isaac Newton's theoretical formula.

Studying Isaac Newton's accomplishments leads to two significant observations. One is the exceptional ingenuity for exploring the unknown. Another one is the devastating limitations created by the paradigm phenomenon. It proves that the human brain is extremely powerful and can follow logical-mathematical thinking paths toward extraordinary innovations. Unfortunately, the same human brain can get caught in an unproductive loop of false thinking and may not find a way out of it.

Isaac Newton not only developed a precise formulation of the law of gravity applicable to anybody in the universe, but he also cracked open the dome of heaven, arguing for the infinity of space. [14]

Albert Einstein's Search for Logical-Mathematical Intelligence of Relativity

Albert Einstein (1879–1955), a German-born theoretical physicist, started his professional career at the patent office in Munich, Germany. Being very efficient in his official daily chores, he had time to follow his personal interests. He felt puzzled by the concept that a person's judgment of movement depended on his reference point. His work and the results achieved turned the world upside down. [23, 49]

Being located on Earth, the moon seems to be moving across the nightly sky. Being located on the moon, one would sense that the Earth is rotating

Figure 4.6 Albert Einstein (1879-1955)   (Internet)

around the moon. Further, sitting on a moving train and observing an oncoming train moving at the same velocity gave Einstein the notion as if the oncoming train was moving at twice its actual velocity. On the other hand, it had become known that, independent of the observer's reference point, the velocity of light would always appear to travel at the actual speed of light.

Einstein felt the urge to explore this circumstance of relativity and delivered his theory of relativity in 1905. The process of reaching a meaningful result was an arduous mathematical task. Fortunately, his work resulted in a transparent yet very powerful mathematical expression for the distortion of time as the velocity of a body increases to the velocity of light. Einstein's Special Relativity Formula can be understood by anyone with a basic appreciation of mathematical square root operations

$$ y \quad = \quad 1 / \sqrt{1 - (v/c)^2} $$

where (y) is a factor that marks the distortion of the progression of time for a body that travels at the velocity (v). When (v) approaches the speed of light (c), (y) approaches infinity, meaning that the time progression of time becomes infinitesimally small and eventually the time will stand still completely.

The charts in Figures 4.7 a and b illustrate the time distortion (y) for velocity (v) shown in percent of the speed of light. Up to 35% of the speed of light, the distortion remains unnoticeable. 35% of the speed of light would be equivalent to approximately 67,000 miles per second, a very impractical velocity. Thus, the classical Newtonian mechanics, which specify velocity calculations as

$$ \text{Velocity} \quad = \quad \text{Distance} / \text{Time} $$

cannot be exactly correct although it is a very practical approach for our common applications on Earth. Under strictly theoretical considerations, it is not absolutely correct.

Once our super-theoretical traveler reaches a velocity equivalent to 85% of the speed of light, the time distortion has reached the factor 2, meaning that the time progresses at half its normal progression on Earth. From then on, though, the distortion increases steeply and reaches the factor 8 at 99% of the speed of light. At a speed equivalent to 100% of the speed of light, the time distortion will approach infinity, i.e., time will stand still completely.

When Einstein presented his Special Relativity to the science community in 1905, there was no possibility yet to prove the concept by way of a practical experiment. However, the status of science in general had reached the point, where Einstein's verbal interpretations and his analytical approach were received with acknowledgement, although the missing experimental proof had always caused certain doubts.

The ultimate proof offered itself when very accurate atomic clocks were developed using the absolutely consistent decay rate of radioactive materials as cadence provider.

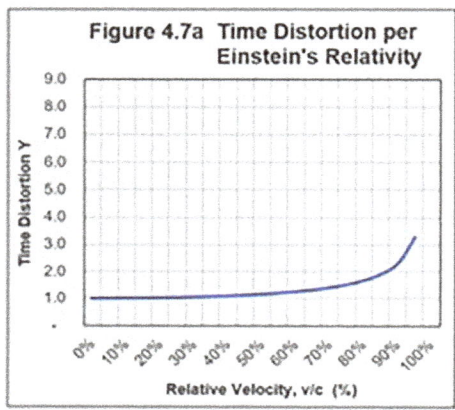

Figure 4.7a Time Distortion per Einstein's Relativity

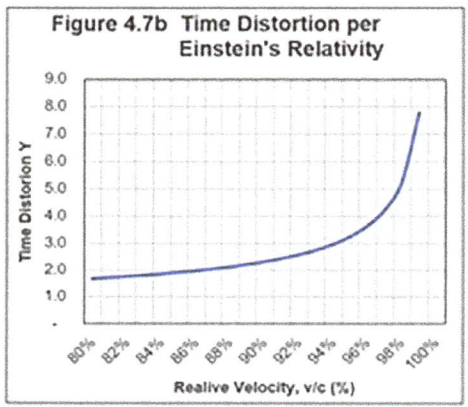

Figure 4.7b Time Distortion per Einstein's Relativity

NASA's International Space Station (ISS) was selected as the host of one of a set of completely identical twins of atomic clocks. After one of the atomic twin clocks had traveled for six months at a velocity of 20,000 miles per hour, the time displays of the twin clocks were compared. The one that had been on the ISS showed a time distortion equivalent to Einstein's Special Relativity Formula.

The experiment not only proved Einstein's Relativity formula to be correct, but it also revealed that not only does living organic matter experience time distortion, but non-living inorganic matter like radioactive materials also experiences the natural law of relativity. It indicates that if any time dependent processes move at the speed of light, their processes will be standing still. Einstein's relativity offers breakthrough insight into astrophysics and our beginning understanding of the universe.

Newton's and Einstein's disciplined applications of logical-mathematical intelligence, and the extraordinary results achieved by both of them, illustrate the exceptional power of disciplined LMI. Both accomplished their findings without today's powerful computational capacity and without today's AI.

With the exceptionally powerful platforms available to everyone in the digital age, LMI can be practiced by students, business professionals, and by curious members of the general public.

Chapter 5
The Nature of Logical Mathematical Intelligence

Introduction

The concepts of Zero and Infinity have attracted human curiosity since ancient times.[22] Zero and Infinity represent concepts that do not fit easily into the number system and have long existed only as concepts with varying definitions. Since approximately 3,000 BC, the phenomena have been mentioned as concepts by philosophers until mathematics started to blossom and added mathematical interpretations to the phenomena of Zero and Infinity.

Understanding the meaning of Zero and Infinity is very important in today's industrial age and its societal systems. Many of the systems that affect our society's industry, economy, and our personal lives, are inherently unstable and can derail unless certain rules are imposed on systems and are consciously maintained by policymakers, captains of industry, and our citizens. In some cases, instabilities observed in certain economic and physical systems are closely tied to the concepts of Zero and Infinity and the combination of both.

Using the concepts of Zero and Infinity and gaining a fundamental understanding of their functionalities leads to a surprising simplicity of the inherent instabilities of systems that are applied to manage business entities, industries, and economics, such as growth, price inflation, unemployment, and financial transactions.

Zero as Concept and Numeric Entity

The concept of "nothing" gave ancient philosophers great difficulty. They asked, "How can nothing be?" The first realistic view of Zero came into being when the concept was included as a space holder in counting systems. Zero as a space holder became a fundamental component of the Hindu-Arabic numeral system, which used the base 10, and was brought to Western Europe in the 11th century AD. The original system included the figures 9, 8, 7, 6, 5, 4, 3, 2, and 1, but not yet the figure zero. The practicality of the Hindu-Arabic System, which marks the beginning of the decimal system, contributed to its survival. The progress in mathematics was slow, because the Roman System, which used letters and did not include a space holder for "nothing", was applied concurrently and prevented the use and acceptance of zero as a numeral figure.

The concept of Zero as a numeral figure that plays a role in calculus remained in Western Europe unknown and was considered dubious for centuries. During the dark Medieval Ages, when scientific knowledge was denied in Europe, the definition of Zero became a contentious subject. People who felt that they could provide answers to the question of what Zero means were thought to be inspired by the devil and were treated like witches who deserved to die. This circumstance contributed to a hold on progress in mathematics until Europe grew out of the Medieval Ages and rediscovered the knowledge of ancient Greece and the societal benefits of science.

Zero became known as the lowest integer, which had its position on the number scale right at the transition from the negative scale to the positive scale. The transition from Zero-as-a-concept to Zero-as-a-numeral-figure was further confirmed by the mathematical process of division. Division can provide a remaining balance, which is a mathematical entity, or the remaining balance can amount to "0", which must be a mathematical entity, too.

The next logical step was the exploration of the multiplication process. Here the combined understanding of Zero as a concept and as a numeral figure was very helpful. Any number being multiplied with the number of zero amounts to zero, which makes good logical sense. Consider that a number represents the value of an object, and multiplying that value with zero essentially means that the value does not exist. Thus, multiplying a number by zero must amount to "0".

The process of division by zero remained an unresolved mystery for a long time. One can divide any number by any number that is larger than zero. But dividing a number by zero does not offer a logical answer. Yet, the LMI that dominates business decision-making processes may involve correlations that do include the possibility of divisions by zero. More specifically, the stability of economic systems may depend on parameters that control the denominator of a division, and we have to make sure that those parameters do not become so small that the system's stability is endangered.

The latter concern leads us to the second phenomenon, the concept of Infinity and the stability and/or instability of important systems of modern societies.

Infinity as a Philosophical Concept and as a Mathematical Phenomenon

The concept of Infinity has interested philosophers since ancient Greece. Infinity has been understood as being boundless. In the 17th century, mathematicians introduced a mathematical symbol for infinity (∞) and explored mathematical operations involving the Infinity concept. [22]

Infinity applies to both directions of a number scale; it can apply to infinitely large positive and negative numbers without bounds. It can also refer to extremely small entities. The latter is referred to as infinitesimal calculus, which deals with infinitely small quantities

Infinity as a boundless concept resulted in practical applications in series calculations, where infinitesimally small process steps are integrated endless times to obtain integrated results. For example, the integration of infinitesimally small linear segments of a circle's circumference yields the factor (π), where $\pi = 3.14$. The factor π results in a simple formula for calculating a circle's circumference "C" using the radius (r). A circle's circumference is determined by the formula $C = 2 * \pi * r$. In a way, this formula is the result of applying the concept of Infinity in a practical manner.

Of particular interest is the interplay between Zero and Infinity. Both seem to be extremely contrary. One represents a single distinct value, i.e., absolute zero, while the other one represents a boundless range of extremely large numbers. However, the Infinity concept also includes the concept of boundlessly small quantities and, thus, in extreme smallness, Infinity's infinitesimal quantity approaches the quantity Zero.

The interplay between Zero and Infinity reflects the stability limits of important systems in our modern society and its economic performance. Understanding this can help us to avoid major societal crises. Consequently, the interplay between the mathematical functionalities of Zero and Infinity must be treated as an important component of the Logical-Mathematical Intelligence (LMI) in the digital age.

In the Section "Zero as Concept and Numeric Entity," it was observed that divisions by zero cause a seemingly unresolvable issue. There is no logical conclusion that can lead to an explanation. In Microsoft's Excel program, the results of occasional inadvertent divisions by zero are simply marked as "#Div/0!" without any further explanation. Mathematical logic can assist in exploring what is happening.

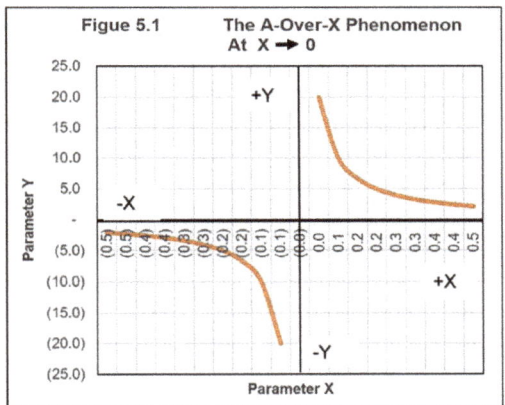

Since a direct explicit division by zero is undefined, mathematics can only assist by applying the division process with an "approach" to zero. This approach is exercised in the charts of Figure 5.1. As a positive number (A), which is set here equivalent to 1 (A = 1), is successively divided by a declining divisor (X), the result for (Y) gets larger and larger. As the divisor reaches extremely small values, the Y-value climbs steeply and approaches infinity. This is typically illustrated by the mathematical expression

$$Y \quad = \quad \frac{A}{X \to 0} \quad \to \quad +\infty$$

71

If we follow the X-value from a large positive number down toward zero, the Y-value represents initially near-zero values, rises sharply, and approaches positive infinity. As we continue the process, reach across the 0-mark on the X-scale and enter into the negative section of the X-scale, the Y-value mysteriously jumps from positive infinity to negative infinity and continues on to approach a 0-value. This effect is represented by the formula:

$$Y \quad = \quad \frac{A}{-X \rightarrow 0} \quad \rightarrow \quad -\infty$$

The above exercise results in an interesting phenomenon that happens at the 0-mark on the X-scale. As the X-values march from positive across the 0-mark to negative, the Y-value first disappears into positive infinity, and then it re-appears from negative infinity. Thus, at the 0-mark on the X-scale, Y can be at positive and at negative infinity. Consequently, at X = 0, the Y-value is not defined. This is why (Y) remains undefined at X = 0. In addition, infinity is not a practical number anyway. Consequently, all considered, if a system results at the condition where a division by "0" applies, the system would oscillate between large positive and large negative control values and, thus, reflect an enormous instability.

For business and societal intelligence considerations, the mathematical characteristics of an A-Over-X function can be important. System functionalities may feature A-over-X relationships, where (A) represents a constant of sorts and (X) may represent a manageable parameter. If the (X) parameter is allowed to approach extremely small values, the function (Y) may raise excessively and drive the entire system into a serious instability.

The described trend toward system instability is practiced by the money flow system and the money multiplier effect, which is described in detail in Chapter 3.

It is interesting to note that the human brain is continuously making speedy assessments of observed objects. This assessment can also include the use of the described A-Over-X function and the (X) parameter can indeed reach the value of zero. This effect is described in the next section. The fact that, under certain conditions, the human brain suffers from the described instability testifies to the natural existence of unstable system conditions independent of mankind's understanding of the logical-mathematical intelligence involved.

Eye and Brain Interactions – An Insight into the Brain's Illusionary Power

Recognition of stationary and moving objects has been known as an essential capability for survival. Extensive research of the different mechanisms involved in eye-brain interactions provides consistent knowledge of the exceptional phenomena that play important roles in vision and vision interpretation processes. The human eye, with its large number of receptors, provides the mechanism for sending still images to the brain. In spite of the eyes extraordinary biological design, it simply performs like a still-image camera that has no built-in intelligence for decision-making capability. The burden of intelligent, accurate, and fast interpretations rests with the human brain.

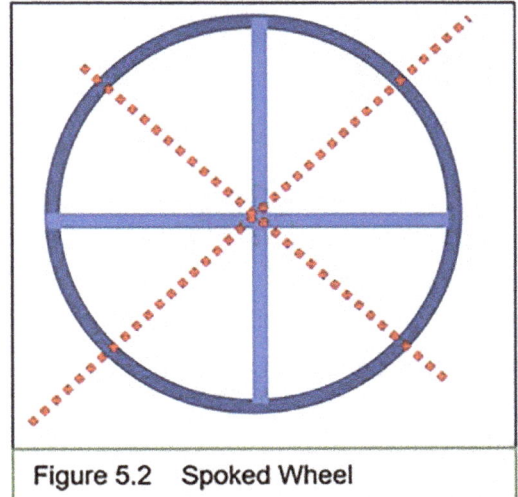

Figure 5.2   Spoked Wheel

When the head and the eye remain stationary, a moving object will successively affect neighboring retina receptors. These effects are interpreted by the brain as a moving object.

Assume now a stationary object and either the eye or the brain or the observer's entire body is moving. The stationary object will successively affect neighboring receptors although the brain will still interpret the object as a stationary object. We may conclude that the brain takes all things into account, i.e., the movement of eyes, the head and the

73

body, and the successive stimuli of neighboring receptors. The brain's integral assessment enables the brain to correctly determine if the object is moving or not. [17]

The human brain performs a phenomenal task, all day long, again and again. Each of the subtasks involved engages mathematical functions, and the brain performs these functions concurrently and arrives at mathematically correct conclusions. The result of all the constantly applied intelligence is what we actually "see." The eyes are taking in still images, one after the other, but what we "see" is what the brain concludes and what the brain wants us to see. Thus, humans see their brains' "illusions." Most of the time, these illusions reflect reality correctly, but sometimes the illusions can be wrong. Interestingly, the wrong conclusions provide some insight into what the brain actually does and they reveal the mathematics applied by the brain. Observing a rotating spoked wagon wheel can provide that information. [2]

Consider a wagon wheel with a number of evenly spaced identical spoke images. The wheel chart shown in Figure 5.2 includes four evenly spaced spokes marked in blue. The midpoints are marked by red dotted lines. As the wheel starts rotating at increasing speed, the observer will experience his brain's illusions that can be traced by logical-mathematical intelligence.

The wheel is rotating at a certain number of rotations per minute (rpm). The human eye sends 20 still images per second to the brain, i.e., the images arrive at the brain at 50-millisecond intervals.

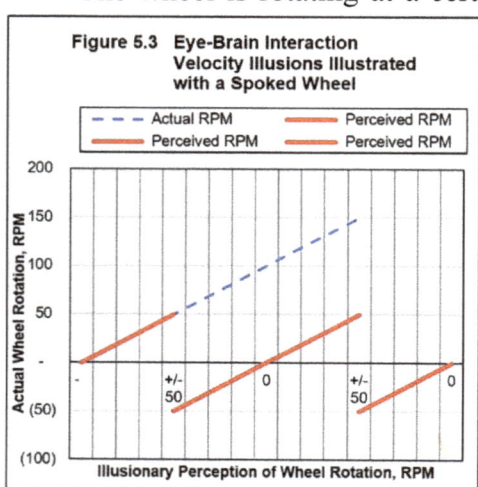

Figure 5.3 Eye-Brain Interaction Velocity Illusions Illustrated with a Spoked Wheel

At successive eye intervals of 50 milliseconds, the brain takes all spokes' positions and compares them with the spokes' positions at the previous eye interval. For this assessment, the brain applies a "neighborhood" analysis. [2] We want to discuss six scenarios. The discussion points are reflected in the diagram of Figure 5.3.

74

1. If, after one full eye interval, the spokes have not yet reached the midpoints, i.e., the spokes' positions are closer to their previous positions than to the next spoke's earlier position, the brain concludes that the spokes have moved forward and, thus, produces an illusionary forward movement for all spokes. In this case, the brain's illusion reflects the reality correctly. In the diagram, the solid red line is consistent with the dotted blue line in the range of 0 to 50 rpm.

2. If the wheel is spinning faster than in Scenario 1 so that all spokes reach the midpoint after one eye interval of 50 milliseconds, the brain finds that the spokes' distances to their earlier positions and to the next spokes' earlier positions are identical. Now the brain's decision is indifferent, and its illusion reflects the critical situation. The observer experiences an unstable rotation that is constantly flipping back and forth between maximum forward and maximum backward rotations. This is illustrated by the chart at 50 rpm, which is the first critical speed of this series of scenarios. In this scenario, since the spokes fall successively on their original and on the midpoint positions, the brain creates the illusion that the number of spokes has doubled.

3. If the wheel's rotational speed is steadily increased further, the brain finds that the spokes are consistently getting closer to the next spokes' previous positions and concludes that the next spokes have moved backward. Now the brain's illusion gives the observer the impression that the wagon wheel is rotating backward. (See the actual speed range from 50 to 100 rpm)

4. As the rotational speed continues to rise, the illusionary backward rotation decreases gradually. When the rotational speed puts all spokes within one eye interval exactly on the next spokes' previous positions, the brain concludes that the wheel is standing still. (At 100 rpm for the actual wheel rotation)

5. Following the illusionary stand still reached in Scenario 4, the process starts again from the beginning, i.e., the brain reports

an illusionary slow forward rotation although the wheel is actually rotating at high speed. For example, when the illusion indicates a forward speed of 25 rpm, the wheel is actually spinning forward at 125 rpm.

6. According to the chart, the 2nd critical speed occurs at 150 rpm.

The logical-mathematical intelligence exercised by the brain can be written in algorithmic form. The example below uses 12 spokes. The critical midpoint position is doubling the number of inter-spoke spaces, which is reflected by the number 2 in the formula. Note that the eye interval is in milliseconds, which needs to be converted to seconds in the formula.

The Brain's Logical-Mathematical Intelligence

Critical RPM $=$ (sec/min) * 1000 / (eye-interval * # spokes * 2)

Critical RPM $=$ 60 * 1,000 / (50 * 12 * 2)

1st Critical RPM $=$ 50 rpm

Figure 5.4 Human Brain Illusions Illustrated Using a Rotating Spoked Wheel					
Description	First Stand Still	First Rotation Reversal	2nd Stand Still	2nd Rotation Reversal	3rd Stand Still
RPM, Actual	0	50	100	150	200
RPM, Brain Illusion	Actual	+/- Infinity	0	+/- Infinity	0

The wagon wheel phenomena reflecting the eye-brain interactions have been illustrated by an electronic simulation. Dr. Michael Bach, professor at the University in Freiburg, Germany graciously agreed to have the model's URL referenced here. [2]   The reader may exercise Dr. Bach's model and experience personally his/her brain's powerful illusions.

The diagram that illustrates the critical wheel rotation situations, where the brain is confronted with constant jumps between maximum forward and maximum backward rotations, reveals a close relationship to the often-observed One-Over-X functionality illustrated in Figures 5.1, 5.3, and 5.4. The brain's analytical features and its illusionary power testifies to the natural validity of the LMI.

In this digital age, where most systems analyses are performed electronically at the highest possible speeds and where simulations are being added to illustrate dynamic situations, the simulation developer needs to be aware of the human eye-brain interactions and the brain's power to constantly turn the observed situational dynamics into illusions. The eye works as a still-image camera, not as a video camera, and all the dynamics that we humans think we see around us are illusions created by the brain. Most of the time, the illusions reflect reality correctly, but for critical situations, the brain's illusion may be incorrect.

The brains illusion can be observed by observing the wheels of adjacent vehicles on the road. At certain velocities, and depending on the number of spokes on the wheel, one can gain the impression that a wheel is turning backward.

At the airport, looking from the front into a starting or shutting-down turbo jet engine, one can observe the wagon-wheel effect. The turbo-blades of the engine's compressor are identical and are equally spaced. At the right rotational speeds, the compressor turns forward, creates a fuzzy image at its critical speed, slows down, stands still, and reverses its direction.

Wherever equally spaced identical objects are moving, the eye-brain interaction can cause the brain to create illusions that don't reflect the reality.

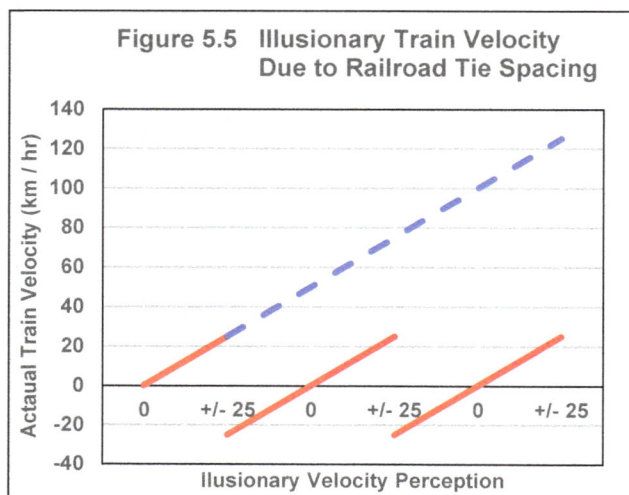

Figure 5.5   Illusionary Train Velocity Due to Railroad Tie Spacing

Ilusionary Velocity Perception

The illusions' deviation from reality can be severe. In the analysis of complex systems, the illusions may guide an observer in the wrong and possibly fateful direction. When pilots of passenger and freight trains misjudge their trains' speeds and enter curves too fast, the identical and equally spaced railroad ties can give them a false perception of speed. As shown in Figure 5.5, at the critical train velocity of 25 km per hour, the number of railroad ties seems to have doubled, and the perceived velocity jumps back and force between +25 and -25 km per hr. In the positive range, the perceived train velocity can be off by a factor of 3 and more. For example, at actual speeds of 70 km per hour, the perceived speed may amount to approximately 20 km per hour.

In this scenario, in contrast to the wagon wheel experiment, the equally spaced railroad ties are actually standing still, and the observer is moving at a certain velocity. This reversal in conditions does not matter. As Einstein noted, the relative velocity counts, which applies to the brain as well and creates its illusion. Catastrophic accidents are said to happen at their own logic. The brain's illusion can certainly contribute to that a fateful logic.

The brain's illusionary power is guiding us humans all the time and gives us a true and correct understanding of the world around us. However, as proven by thought experiments involving the wagon wheel and the railroad ties, the brain's illusions can also be very misleading and can even lead to mind blocking paradigms.

Personal Knowledge Management and The Dynamic Knowledge Matrix

Effective personal knowledge management (PKM) is important for creating sustainable worker productivity. PKM is the process of collecting information, including gathering, classifying, storing, searching, retrieving, and sharing knowledge. The development of PKM principles

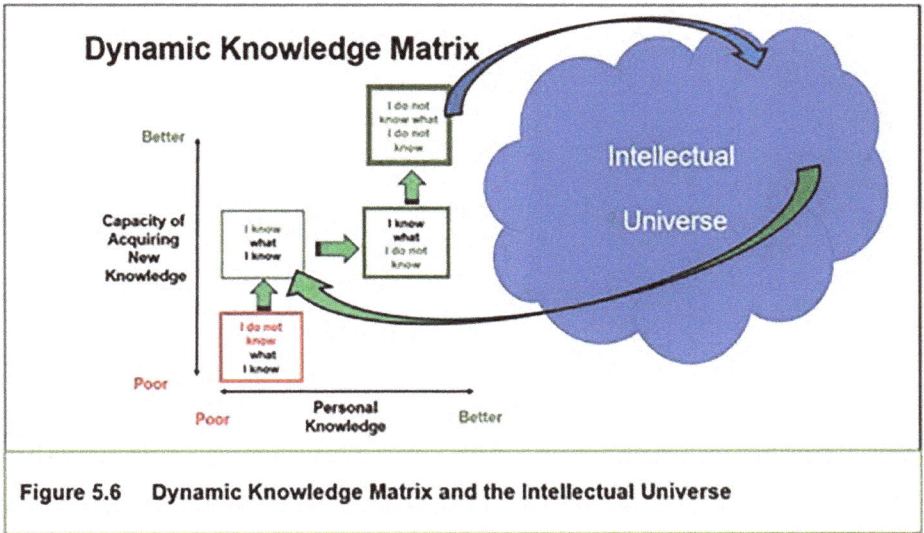

Figure 5.6   Dynamic Knowledge Matrix and the Intellectual Universe

is a response to the idea that workers need to be responsible for their own growth and learning. It is a bottom-up approach to knowledge management (KM).

Effective knowledge management at all levels within an enterprise has become very important. In modern flat enterprises, innovation does not only reside in the minds of brilliant scientists anymore. It can blossom almost anywhere in organizations that are properly structured and provide a culture that makes invention flourish. [25] Consequently, it is critically important to spread the concept of logical-mathematical intelligence and knowledge to groups that are expected to foster innovations.

It is equally important to spread basic concepts of knowledge management and to emphasize the importance of knowledge acquisition, retention, and retrieval. Further, it is important for workers at all levels to attain the skills to access the source of exceptional ideas and solutions.

We call that source the Intellectual Universe (IU). The novel dynamic knowledge matrix can be a guide toward that goal.

The 44th US President, Barack Obama, was once asked what he perceived being the most difficult part of the presidency. He answered, "The most difficult situations occur when I become aware that 'I do not know what I do not know'." [35] One has to pause and read the statement several times to grab its deep and meaningful context.

The quote includes two statements, each one can have a positive and a negative version, which makes four statements illustrated in the dynamic knowledge matrix.

1. At the lowest level of personal knowledge management is the statement "I do not know what I know." This category resembles a very low knowledge management situation. A person who can identify the status of his personal knowledge management with this state must be guided to move over to the next level.

2. At the second level of personal knowledge management, we experience a vertical movement, i.e., the personal knowledge is not expanded, but it is strengthened and brought into that person's sphere of awareness. This second state is indicated by the very strong and positive statement, "I know what I know." This level reflects a very firm awareness of previously acquired knowledge and intelligence. It also reflects good knowledge retention and retrieval capability. Employees with this perception of knowledge and intelligence are most valuable for the enterprise.

3. At the third level of personal knowledge management, we experience a horizontal movement to greater level of curiosity toward new knowledge, indicated by the statement, "I know what I do not know." This awareness level is typically accompanied by an urge to explore the "I do not know" part of the realm of intelligence.

4. At the fourth level of personal knowledge management, we reach into the ultimate crisis level of intellectual awareness: "I

80

do not know what I do not know." However, this must not be perceived as a crisis without an outcome. In fact, this stage of personal knowledge management does not reflect a rare occasion. It is the launch pad toward new ideas. We must tell our employees and our business management students that this fourth state is not the end of the road and that they should not be scared by the observation that there is no solution in sight. Instead, this is a glorious moment and can be the beginning of something great. [28]

5. The fifth state of personal knowledge management reaches into what we call the "Intellectual Universe," which offers an unlimited abundance of solutions.

6. The art is to recognize them and know how to pull them into our world of realities. Reaching into the Intellectual Universe enables us to look nature into her eyes and helps us to spot solutions. Nature's solutions typically reflect surprising transparency and simplicity.

History has shown us how effective nature's transparent and simple solutions are. We do not have to be a Newton or an Einstein to find these winning solutions, but we need to be in command of some logical-mathematical intelligence to be able to build some logical-mathematical knowledge. Even a low-level framework of LMI and LMK can be a key for opening the path into the Intellectual Universe.

Linking Concepts and Innovations

The previous chapter illustrates the path toward the solution-rich Intellectual Universe. There are effective techniques on how to scan the Intellectual Universe for new ideas, new products, and new processes.

A very effective approach for scanning one's Intellectual Universe for new products and services is to integrate two known entities. One entity to consider is the rich family of known and proven concepts. An example is the hub-and-spoke concept. It first presented itself in the form of the money-based market system with its striking advantages over the primitive barter system. The advantages of the hub-and-spoke concept are clearly illustrated by a simple set of LMI, which is an example of a very basic and simple confirmation of an overwhelmingly powerful idea.

The hub-and-spoke system is in itself so powerful that people implement it wherever they can without being aware of what their brain is actually accomplishing for them. Consequently, the hub-and-spoke systems do not have to be consciously implemented; they just happen to emerge by themselves and have existed throughout history.

A historic application of hub-and-spoke occurred when the Central Library of Alexandria was built. The Library of Alexandria became the hub for storing knowledge and creating new knowledge; the hub's spokes were the collection of knowledge from all over the world and the distribution of new knowledge across the world. The idea of a central library was born out of Aristotle's and Alexander's thought experiment with the aim to enhance knowledge and education for the benefit of the world. The underlying hub-and-spoke system, a blessing in disguise, supports the idea superbly and became a natural ingredient of education systems. Today's universities practice the same concept.

As new technologies emerge, concepts like the hub-and-spoke concept can be applied to functions in our society that were theretofore excluded from hub applications. One example is the vast mail distribution network. The existence of high-speed jet airplanes makes the hub-and-spoke concept a viable infrastructure for postal services. Another example is the network of interconnected personal computers, which functions as a super hub-and-spoke system for knowledge collection and distribution.

When known concepts are innovatively applied in new ways as new technologies emerge, the innovation process offers itself like a natural advancement within society. A more complex innovation process occurs when an entrepreneur is inspired by a pioneering thought experiment although the technology required for its implementation does not exist yet. A striking example for this kind of innovation are the disciplined inventions of time-division and code-division multiple access technologies required to support the pioneering thought experiment, which envisioned concurrent wireless voice and data transmissions for a multimillion-user community.

A third group of inventions involves theoretical work toward new logical-mathematical intelligence. These inventions are born out of inventors' inquisitive scientific minds. They are curious about nature, and their work discloses nature's hidden laws in physics and chemistry. They use mathematics as the language to communicate with nature. The LMI is expanded concurrent with the expansion of human intelligence in the natural sciences. Examples for this type of inventions are provided by Pythagoras, Euclid, Newton, Leibnitz, Gauss, Einstein, Heisenberg, and many more.

Chapter 6
Logical-Mathematical Intelligence in the Digital Age

Introduction

Since Adam Smith's emphasis on the division of labor being a major force enhancing the industry's productivity and the nation's wealth, Smith's concept has penetrated all organizations and has created specialization wherever possible. The result has been flourishing economies around the globe. Science, education, industry, and governments have practiced the concept and have achieved the results envisioned by the founder of modern economics 250 years ago. In the digital age, high-speed computing, AI, machine learning, and IoT are changing the landscape and are reversing selected aspects of the division-of-labor concept.

Digital-age technologies combined with high-speed computing can enable comprehensive work process integrations. While the computer is still treating workflow in a step-by-step manner according to the division-of-labor concept, it is resolving each step at high speed, and it accomplishes the step-to-step interface without any delay. Thus, the division of labor, known as a time-reducing and productivity-enhancing methodology, is taken to the next higher level in terms of reducing time and enhancing productivity.

In mass manufacturing, robots are performing individual work elements at high speed and with great precision, such that the individual steps are merged into one continuous flow of a single production process. Consequently, the modern labor force has to think and act now in terms of the entire integrated process.

A striking example of the transition from division of labor to integrated systems is being exercised by Tesla Motors. Traditional automobile manufacturers outsource approximately 80% of the design and manufacture of vehicle components to specialized manufacturers and limit their own activities to the assembly of their vehicles. This procedure symbolizes Smith's division of labor. In contrast, Tesla Motors organized the manufacture of all components within its own organization. The objective is to remain fully in control of the design, manufacture, and quality of all components, thus creating a direct iteration cycle for optimizing the components' function, quality, and performance. [24,25,51] One must add, though, that traditional combustion-driven vehicles entail approximately 3,000 parts while electric vehicles contain only 300 parts. This makes it so much easier for electric vehicle manufacturers to accomplish advanced integration.

Thoughts on AI[3]

Artificial Intelligence, sometimes referred to by the more logical term machine intelligence (MI), is a powerful tool to address problems in a number of sciences that traditional computational approaches cannot resolve. AI assists in how we envision and design safe and resilient infrastructures for today that will be especially functional and effective for society in the future. It enables virtual infrastructures that will remain hidden from our conscious world, and they will have far reaching impacts on how we will work and live. [32, 52]

AI has the potential to greatly enhance applications of logical-mathematical intelligence and digital technologies by leveraging its capabilities in data analysis, pattern recognition, and algorithmic processing. Through machine learning algorithms and advanced analytics, AI systems can process vast amounts of data and identify patterns, correlations, and insights that may not be immediately apparent to human observers.

In the realm of logical reasoning, AI can aid in complex problem solving by employing logical algorithms and rule-based systems. It can analyze intricate logical structures, identify logical fallacies, and even assist in generating logical arguments. Additionally, AI-powered systems

---

[3] Dr. Cory Scott, Professor at Alliant International University, contributed to this Section

can contribute to automated theorem assisting in the verification and validation of mathematical proofs.

AI technologies, such as natural language processing and machine translation, enable the extraction and understanding of mathematical concepts from various textual resources, making mathematical knowledge more accessible and facilitating cross-disciplinary collaborations.

AI plays a crucial role in various areas such as high-performance computer systems and the Internet of Things. There are emerging business needs that AI can fulfill, and we should focus on those. Project management, for example, is an area where AI can have a significant impact, improving efficiency, decision-making, and resource allocation.

Another area to be mentioned is data analytics. In the era of data-driven decision-making, AI plays a pivotal role in transforming the data interpretation process and how businesses operate, strategize, and grow. AI can accurately predict future trends and deliver insights that propel businesses forward.

AI algorithms have the capacity to sift through immense volumes of data, correctly identifying patterns, trends, and correlations that can be easily overlooked otherwise. These algorithms turn volumes of raw data into actionable insights swiftly, enabling businesses to make informed decisions in real time. In a world where data is the new oil, AI-based analytics serve as highly efficient tools, turning raw data into the fuel that drives business growth.

AI-powered predictive models are constantly learning and evolving. They enhance the accuracy of their predictions over time by learning from new data. These models can accurately forecast market trends, demands, and consumer behaviors. This level of foresight allows businesses to stay proactive and make data-driven decisions that align with future trends.

It's important to recognize that while AI can unlock the full potential of systems, it can also reveal latent instabilities. AI's application in financial systems and its potential implications in connection with the emerging cryptocurrency have not been addressed but seem important. We should not learn from cryptocurrency-imposed economic failures but utilize AI-based analytics to identify ahead of time how the economy may react and how failures can be prevented. It is crucial to analyze the potential risk and to ensure that societal impacts are positive.

While AI holds promise in enhancing logical-mathematical intelligence, it's important to recognize the limitations. AI systems are dependent on the quality and representativeness of the data they are trained on, and biases present in the data can potentially influence the outcomes. Additionally, the interpretability of AI models in logical reasoning can be a challenge, as their decision-making processes may not be readily explainable.

AI doesn't necessarily require the analyst to be a computer scientist. Rather, it's about being able to analyze and explore the various aspects of systems affected by AI, such as financial systems. Applying logical-mathematical intelligence can help us gain insights into a system's functionalities and potential risks.

In summary, it becomes obvious that AI is the perfect technology enabling the effective integration of logical-mathematical intelligence into the Synergetic Five-Step Innovation Process.

Digital Transformation

Innovation, the first major step toward improving productivity, is the cornerstone of successful enterprises, not only today, but throughout history. In the age of digital transformation, forward-thinking enterprises embrace new trends and cultivate for the ongoing enterprise a continuing start-up mentality. [51] Leading corporations are teaming up with leading university-based research teams to embrace all aspects of the digital transformation trend.

With thousands of employees and a virtually infinite number of business operations, processes in large corporations are highly complex and pose an enormous challenge for business analysts and business developers. The solution to this challenge lies in advanced digital technologies, i.e., artificial-technology-based tools, such as process mining and digital twin.

Process mining takes the well-known data mining methodology to another level. While data mining extracts from data stores specific data defined by attributes and properties, process mining extracts complete data series that belong to specific business processes. Process mining is a family of techniques relating the fields of data science and process management to support the analysis of operational processes based on event logs. Process mining is a revolutionary new data analysis technology that allows business processes to be mapped automatically in a transparent manner that enables business analysts to reconstruct the process in any desired variation. At the push of a button, the user obtains a snapshot of the entire process, which enables him to identify redundant process steps, cost drivers, weak points, and bottlenecks.

Process mining is not just an effective tool for business analysts and developers with the aim to improve the enterprise's productivity; it has also emerged as an effective methodology to perform business and financial audits.

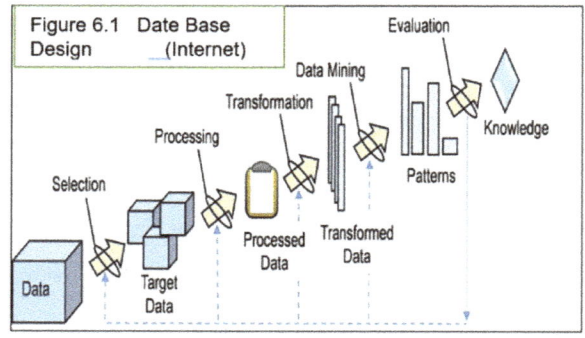

A pre-requirement for successful process mining is that the database is set up properly. If one wants to combine data from two (or more) different IT systems, the most important thing to look for is how one can follow a case across these different systems. For example, if the Enterprise Resource Planning system (ERP) uses a purchase order number for each case, and the Financial Management System (FMS) uses an invoice number for each case, then one needs to find a back reference in the invoice to which purchase order it refers to. This back reference is accomplished in the processing and transformation phases indicated in the process flow diagram of Figure 6.1.

Process mining applications can be used in specific focus areas to run highly detailed reviews within the process, such as where work is being duplicated or unnecessary actions are executed, where delays are increasing throughput times, and how interfaces to suppliers can be improved to achieve maximum efficiency in terms of data consistency, lean process management and process transparency.

Process mining is capable of making processes transparent within the company as well as, like an x-ray image, revealing potential areas of improvement. This new technology provides business managers with active support in optimizing enterprise processes to achieve and maintain exceptional competitive advantage in today's global marketplace. [40]

Digital twin is a term used for digitally automated processes. The term digital twin signals a set of automation applications that utilize digital technologies to simplify and accelerate automation processes with the objective to create visual illustrations of work processes.

In product and business development environments, a digital twin can be viewed as a common reference model. The first step would be the data provision, i.e., the identification and extraction of relevant digital process activities. For this purpose, minute digital process elements are extracted from more of the enterprise's IT systems, for example, as shown in the display, a 3-D Computer Aided Design (3-D CAD) tool. Each relevant event booked within a local system is automatically transferred into a central data store.

Figure 6.2 Digital Twin Application Using 3-D Computer Aided Design, 3-D CAT   (Internet)

Subsequently, the data is standardized, i.e., the data elements from the different source systems are unified and harmonized. This enables the durable linkage of logistic process steps of internal and distribution systems. The process modeling facilitates intuitive visualization of the actual processes in the form of digital twins for the experts of corresponding subject divisions.

Process mining and digital twin offer exciting perspectives. In combination with Industrial Internet of Things, exciting opportunities offer themselves. Individual products become increasingly "intelligent" and lend themselves to optimization. Because of normalized data and dynamic platforms, the concept of intelligent twins can be freely expanded. With this digital tool, process experts and data analysts have completely new opportunities for gaining detailed insight for immediately improving process flows.

A digital twin is a digital model of an intended or actual real-world physical product, system, or process that serves as the effectively indistinguishable digital counterpart of it for practical purposes, such as simulation, integration, testing, monitoring, and maintenance.

The digital twin has been intended from its initial introduction to be the underlying premise for product lifecycle management and exists throughout the entire lifecycle of the physical entity it represents. A digital twin of an existing entity may be used in real-time and regularly synchronized with the corresponding physical system.

A digital process twin can be a dynamic software model, which extracts elements of individual activities from IT management systems and visualizes them. This enables management to obtain a simplified overview of a process and to exercise operational control opportunities.

The digital twin can and does often exist before there is a physical entity. This is true for product

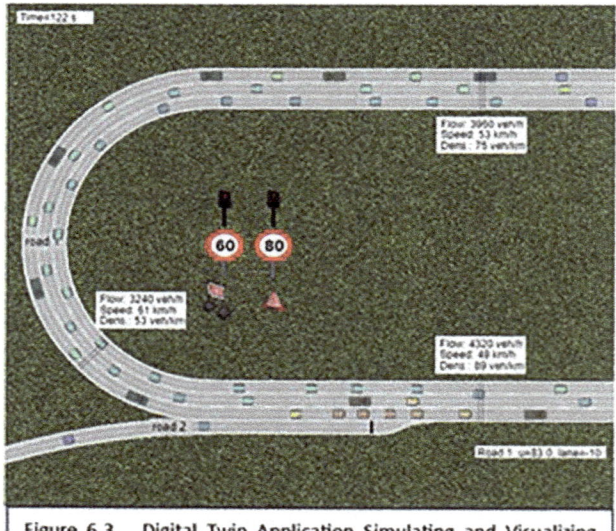

Figure 6.3   Digital Twin Application Simulating and Visualizing
Traffic Flow Dynamics   (Internet)

development using 3-D CAT systems as shown above. It is also true for dynamic model development. In this regard, the author developed the algorithms for controlling traffic flow by way of virtual infrastructures and intends to create digital twins for illustrating to government authorities the intended traffic flow performance on freeways and county roads, as described in Chapter 11.

The digital transformation foremost entails the digitalization of processes. This requires that conceptual descriptions and qualitative assessments must be converted to numeric definitions. For that purpose, existing logical-mathematical intelligence must be identified or established.

Product and process digitization creates the need for major transformations in the market place. The customers' comfort with smart and highly digitalized products and services must be created concurrently. Customer transformation is a challenge that is being addressed by the industry. [28,32,40] As the traditional automobile transitions toward a computer on wheels, the challenge for the general public can be huge. This challenge will be further exacerbated as the traffic flow will be structured and managed by automated traffic management systems. How driver comfort may be managed in the future is analyzed in great detail in Chapter 11.

Virtual Infrastructures and Enterprise Architectures

Virtual infrastructures function the same way as physical resources and facilitate how IT can be effectively used to allocate virtual resources quickly and across multiple systems. Virtual infrastructures can help organizations achieve greater IT resource utilization, flexibility, scalability, and cost savings. [11, 57]

Virtual infrastructures have the potential to add significant efficiencies and cost savings. Physical infrastructures take up a large amount of space, while virtual infrastructures superimposed on physical infrastructures can raise the physical infrastructures' productivity, thus eliminating the need for physical expansion.

Virtual Infrastructure platforms that have been developed are offered in the form of Software as a Service (SAAS) to help companies collect, visualize, analyze, and communicate information to better plan and adapt to change.

Virtualization enhances every aspect of an enterprise's performance. The benefits of a virtual infrastructure include:

- Cost savings: Virtualization reduces capital and operating costs.

- Flexibility: A virtual infrastructure enhances an organization's flexibility and enables it to respond quickly to changes in customer orders and market needs.

- Productivity: IT teams can be more efficient and can respond faster to employee's demands for new tools and technologies.

- Competitive Advantage: With a virtual infrastructure installed, the enterprise can react faster to market changes and, thus, maintain and enhance its competitive advantage.

Emerging digital technologies accelerate the industry's digital transformation, and collaboration between business management and information management has become a critical component for successful enterprises. In this context, virtual infrastructures form the basis of emerging enterprise architectures (EA), which integrate organizations' strategic goals with corresponding technical initiatives. [45] Getting started with EA in the modern digital age is a smart move. Its measurable impacts benefit financial, security, operations, and decision-making outcomes, as well as reducing complexity and facilitating growth.

Technology-integrating virtual infrastructures and enterprise architectures combined are powerful management concepts that yield business transparency and streamline decision-making processes.

Virtual infrastructures by themselves have a major impact on the configuration of selected industry systems and projects within the economy. An effective application of virtual infrastructures is accomplished by superimposing it on existing physical infrastructures to raise their productivity without limits, depending on the status of the digital technologies engaged. See Chapter 11 for elaborate examples.

Digital Marketing[4]

Marketing management has undergone a significant transformation with the advent of AI and advanced analytical techniques. AI has the potential to revolutionize marketing by augmenting logical-mathematical intelligence along with MI algorithms and predictive modeling

I. Defining the Problem and Setting Objectives

Let's revisit the hypothetical example of a premium electronic gadget manufacturers aiming to expand their market share among tech-savvy Millennials. With the integration of AI, marketing teams can refine the problem statement and establish more precise objectives. For instance, the objective could be to increase the conversion rate among Millennials by 20% within the next six months.

II. Data Collection and AI-driven Analysis

To leverage the power of AI, marketing teams need to collect relevant data from diverse sources, including customer interactions, social media, website analytics, and purchasing history. AI algorithms can then be applied to analyze the data and extract meaningful insights.

One AI technique that complements LMI is natural language processing (NLP). NLP algorithms can process unstructured data, such as customer reviews or social media posts, to gain sentiment analysis and extract valuable information about product preferences, features, and pain points.

III. Segmentation and Targeting with AI

Traditional segmentation techniques, when combined with AI, can lead to more refined and personalized targeting strategies. AI algorithms, such as clustering or classification models, can automatically identify micro-segments within the Millennial demographics based on a wide range of variables, including behavior patterns and interests.

---

[4] Dr. Reza Hosseini, Hansa Tek Netics, Vice President of Digital Marketing, Contributed to this chapter

For example, a company may utilize a clustering algorithm, such as k-means, to identify distinct groups of Millennials based on their online behavior, purchase history, and product preferences. By analyzing the clusters, marketers can uncover unique characteristics of each segment and develop tailored marketing strategies to address their specific needs.

IV. Positioning and Differentiation with AI

AI can enhance the positioning and differentiation process by uncovering hidden patterns and relationships within vast datasets. Collaborative filtering, a common AI technique, can be employed to recommend personalized product offerings to individual customers based on their preferences and browsing history.

To illustrate the impact of AI, let's consider a formula commonly used in collaborative filtering called Cosine Similarity:

$$\text{Cosine Similarity } (A, B) = (A \cdot B) / (\|A\| * \|B\|)$$

In this formula, A and B represent vectors representing customer preferences, and $(A \cdot B)$ denotes the dot product of these vectors. $\|A\|$ and $\|B\|$ represent the Euclidean norms of the respective vectors. By calculating the cosine similarity between customers, marketers can recommend products to new customers based on the preferences of similar customers, thereby enhancing the positioning and differentiation strategies.

V. Marketing Mix Optimization with AI

AI plays a pivotal role in optimizing the marketing mix by leveraging predictive modeling and reinforcement learning algorithms. These algorithms analyze historical data, customer behavior patterns, and external factors to determine the optimal combination of product, price, place, and promotion.

For example, a reinforcement learning algorithm can learn from past campaigns' performance data and simulate various scenarios to identify the most effective marketing mix. By maximizing a defined reward function, the algorithm can suggest optimal decisions, such as the appropriate budget allocation for different marketing channels or the ideal pricing strategy.

VI. Performance Monitoring and Decision Making with AI

AI excels in real-time performance monitoring and decision-making, empowering marketers to make data-driven decisions swiftly. AI-powered dashboards consolidate and visualize key performance indicators, providing insights into campaign performance, customer behavior, and revenue trends.

For instance, anomaly detection algorithms can flag unexpected shifts in customer behavior, such as sudden drops in conversion rates or unusual spikes in customer complaints. These alerts enable marketers to investigate and take corrective actions promptly, such as modifying marketing campaigns, adjusting pricing strategies, or optimizing customer service.

Real-World Example: AI-Powered Recommendation Engine

Let's consider an online retailer that employs an AI-powered recommendation engine to personalize product recommendations for each customer. The engine utilizes collaborative filtering algorithms to identify similar customers and suggest relevant products.

Suppose the retailer observes that a customer, Alice, frequently purchases electronic gadgets and has shown a high affinity for the latest smartphones. Using collaborative filtering, the recommendation engine identifies customers who exhibit similar purchasing patterns and have previously purchased smartphones. The engine calculates the cosine similarity between Alice and these customers, prioritizing the recommendations based on the highest similarity scores.

The formula for cosine similarity, as mentioned earlier, is:

$$\text{Cosine similarity } (A, B) = (A \cdot B) / (\|A\| * \|B\|)$$

In this case, A represents the vector representing Alice's preferences, and B represents the vectors of similar customers. By calculating the

95

cosine similarity, the engine identifies the most similar customers to Alice and recommends smartphones that were popular among those customers but not yet purchased by Alice.

This AI-driven recommendation engine allows the retailer to enhance customer experience, increase customer engagement, and boost sales conversion rates.

In conclusion, the integration of LMI with AI revolutionizes marketing management, enabling businesses to leverage vast amounts of data for more precise decision-making. Through AI-powered techniques such as NLP, clustering, collaborative filtering, and reinforcement learning, marketers can unlock deep insights, personalize marketing efforts, optimize the marketing mix, and monitor performance in real-time.

In the hypothetical example presented above, the synergy between LMI and AI empowered the premium electronic gadget manufacturer to target Millennials more effectively, refine positioning strategies, and optimize marketing campaigns. By leveraging AI and embracing data-driven decision making, businesses can gain a competitive advantage in today's dynamic marketplace, ultimately driving growth, customer satisfaction, and long-term success.

Predictive Analytics in the Digital Age[5]

Staying up to date on data and trends is a vital practice for forward-thinking business Intelligence leaders. [13] Many leaders use trends to benchmark their teams' current skills and abilities. Others are ensuring that their future strategy aligns with the latest technology advancements, including AI and ML. Delivering measurable value to the business is considered a top priority. [13]

An informed decision is only as good as the data that drives decision-making. Ensuring data integrity is vital as misrepresentative data does little good toward being accurately informed. Misinformation

---

[5] Dr. Aaron Wester, Associate Professor at Alliant International University, contributed to this Section.

96

introduces a critical level of bias that negatively warps perception through a distorted reality lens.

In this digital age, collecting data for use in data driven storytelling is an important part of the new norm. However, what all stakeholders across organizations struggle with worldwide is what to do with this vast treasure trove of information that sits at the ready for descriptive or predictive modeling, synthesis, interpretation, and recommended prescriptions. There are plenty digital marketing agencies now that try to help stakeholders make sense of the multitudes of data stored across a dizzying number of agnostic silos and modalities. The real test is ensuring that the data is truly representative, valid, and reliable. After all, with great power comes tremendous responsibility. Information is power, and probability-based predictive prescriptions are powerful, especially when valid and reliable data is used in reliable manners.

Those data managers and analysts that become the caretakers of the data have the daunting task of assuring accuracy of the results. Due diligence is required in the extraction, transformation, and loading processes as well as the modeling processes to affirm that the correct formulas are applied based on the data types. A good data analyst knows that extracted information can be processed to output convincing results. That doesn't mean the results are accurate though. Too few data analysts take the necessary time to validate the data prior to reporting that information in slick dashboards and visualizations with dazzling, alluring, and distracting faux finishes.

Validating the data can be inconvenient yet necessary. Checks must be put into place that may lead to a point where the analyst must firmly put a foot down, take a stand, and boldly proclaim that the data is simply not reliable. A data vetting process may include identifying the type of data, which automatically limits the type of data model that can be applied. A large number of questions must be answered as listed in the Figure 6.4.

All these questions matter where non-distorted predictive forecasting and unapologetically accurate probability-based hypothesis testing is demanded.

Leadership and departmental stakeholders rely on the recommended prescriptions provided to them by analysts who have modeled the data to make informed pivotal business decisions. Often these decisions are tied to substantial resource allocations at tremendous cost with the anticipation of high yielding returns. These leaders and stakeholders typically lack the time and understanding to scrutinize the data results in any meaningful or productive way. The onus, therefore, is on the data analysts to ensure they have conducted their due diligence to vet the data properly and apply the appropriate modeling to generate actionable insights and craft accurate data-driven narratives. Data results are only as good and meaningful as the data fed into the model after all.

Figure 6.4 Typical Questions to be Asked In Support of Predictive Analytics In the Digital Age

- Is the data nominal, ordinal, interval, or ratio?
- Is the dataset continuous or discrete?
- Are there null values?
- What type of distribution is exhibited for variables on a histogram?
- How impactful are the outliers?
- Are there critical outliers that can skew the central tendency?
- Do variables exhibit monotonicity and homoskedasticity?
- Do data values tend towards the mean?
- Is homogeneity of variance exhibited?
- Is the data descriptive or inferential?
- Is the data parametric or non-parametric?
- Is the sample size accurate in representing the total population parameter?

Organizations cannot afford to be blindsided by false data-driven narratives. To gain strategic competitive advantage, reliable and meaningful prescriptions are both crucial and essential as leaders must be decisive where winning new business and revenue is necessary for growth and profitability. Level setting and collaboration is key in establishing a cross-functional approach between data scientists and business leaders to ensure the tightest possible alignment. When there is a culture of data literacy, and a clear mutually shared understanding of the business objectives, goals, and challenges across these groups, informed analysts and business strategists can tailor insights and recommendations to help answer key questions such as, "How do these insights impact our marketing strategy?" and "What changes should we make in our product offerings based on this data?"

In the era of data-driven decision-making, AI plays an integral role in transforming the data interpretation process and how businesses operate, strategize, and grow. AI can accurately predict future trends and deliver insights that propel businesses forward.

It's important to recognize that while AI can unlock the full potential of systems, it can also reveal latent instabilities. AI algorithms have the capacity to sift through immense volumes of data, correctly identifying patterns, trends, and correlations that can be easily overlooked otherwise. These algorithms turn volumes of raw data into actionable insights swiftly, enabling businesses to make informed decisions in real time.

AI-powered predictive models are constantly learning and evolving. They enhance the accuracy of their predictions over time by learning from new data. These models can accurately forecast market trends, demands, and consumer behaviors. This level of foresight allows businesses to stay proactive and make data-driven decisions that align with future trends.

AI-driven predictive analytics aren't just about forecasting but also play a critical role in identifying untapped business opportunities and potential risks. By spotting emerging trends and market shifts, they provide businesses with the strategic foresight needed to capitalize on favorable conditions and mitigate potential threats

In industries where demand forecasting and inventory management are critical, AI is an indispensable tool. By accurately predicting demand patterns based on historical data and current market conditions, AI enables businesses to optimize inventory levels. This reduces the risk of surplus stock and ensures that products are always available when and where customers need them.

When AI-based approaches are exercised iteratively rather than in a stand-alone way, organizational leaders can trust the data and will find their ability to be nimble and disruptively adaptive to changing market conditions, customer preferences, and emerging trends to be razor-sharp, precise, and timely. They will find themselves in the enviable position of nearly always being in the right place at the right time, not by happenstance or randomly, but through the power of accurate and reliable AI-based predictive analytics and probability-based modeling.

Cybersecurity – What is it Exactly?[6]

In today's high-tech environment, cybersecurity represents and an important technology-based expertise needed to minimize technology-based project failures. Cybersecurity has become synonymous with computers and information technology (IT) or information systems (IS). But what is cybersecurity, what does each mean and include or exclude, and why is cybersecurity so important?

Webster's dictionary defines cybersecurity as "measures taken to protect a computer or computer system (as on the Internet) against unauthorized access or attack" (Merriam-Webster, n.d. [65]). And while the definition of cybersecurity is more easily understood in terms of the protection aspect, there remains a gap in what the extent of cybersecurity and protection is.

To clarify the terms, let's accept that cyber is more related to IT. IT and IS are typically accepted as a set of fields associated with computers, including the hardware (HW) component (computer), the software component (SW), the processing of the information, and its storage (Chandler, Munday, 2011 [63]). With that, cybersecurity tends to focus on protecting computers and networks but, more specifically, protecting data and its integrity.

That statement explains why security is a significant component in protecting an entity's (corporation, business, academic institution, or personal) data from misuse.

The approach needed to protect an IS has to be more holistic by applying its own set of controls that can seek to provide the optimal set of protections across five major areas outlined by the National Institute of Standards and Technology (NIST), specifically NIST SP800 Cybersecurity Framework (CSF). The five areas of concern are:

1. **Identification**: Develop an organizational understanding to manage cybersecurity risk to systems, people, assets, data, and capabilities.

---

[6] John McCready specialized in Cybersecurity and related Standards. His practices include projects for the US Navy.

2. **Protection**: Develop and implement appropriate safeguards to ensure the delivery of critical services.

3. **Detection**: Develop and implement appropriate activities to identify the occurrence of a cybersecurity event.

4. **Response**: Develop and implement appropriate activities to take action regarding a detected cybersecurity incident.

5. **Recovery**: Develop and implement appropriate activities to maintain resilience plans and restore any capabilities or services that were impaired due to a cybersecurity incident.

So, what do the five CSF areas encompass? What are the methods and scope of a cyberattack, and why is this so difficult? To understand that, one has to consider how information and network systems can be attacked. Terms such as attack profile, vectors, and surfaces are related to how an attacker would begin a systematic approach to attacking a target's IT and IS systems. They are known as the Cyber Kill Chain (the Cybersecurity Exchange, n.d. [64]) listed in Figure 6.5.

Figure 6.5   The Cyber Kill Chain [64]

- Reconnaissance
- Weaponization
- Delivery
- Exploitation
- Installation
- Command and control
- Actions on Objectives

With this systematic approach to identifying a vulnerability within an entity's IS/NS, the tools used by an attacker are extensive. According to the Cybersecurity Exchange (n.d. [64]), the top 10 most common types of cyber-based attacks are listed in Figure 6.6.

Each of these attack types carries its own signature and impact, resulting in different effects anywhere from minor disruptions to email traffic up to a total loss of the IS/NS and a full compromise of the data. Blackmail (Ransomware) is often practiced, where money is expected to be paid to regain access to your data.

In summary, cybercriminals, from the curious (to see if they can

Figure 6.6 Most Common Cyber-Attacks [64]

- Malware
- Denial of Services
- Phishing
- Spoofing
- Identify-Based Attacks
- Code Injection Attacks
- Supply Chain Attacks
- Insider Threads
- DNS Tunneling
- IoT-Based attacks

do it – a.k.a., "script kiddies") through criminal organizations (industrial espionage) and nation state actors (Russia, China, North Korea, etc.), are actively scanning systems or sending phishing emails to trick a user into opening a link or file, thereby opening a path to the injection of a malicious file or code.

The IT industry offers technologies and security-focused partnerships, which address protection, resilience and confidence. Special technologies are designed to protect against cyberattacks, and, when an attack occurred, they can facilitate a rapid and efficient recovery of vital workloads. Service provider bring partnerships and IT presence together to solve the biggest security challenge.

Being alert to anomalistic or unrecognized behaviors is the first and best line of defense against cyberattacks. Cybersecurity-based systems' architectures provide the framework to maximize the protection of businesses and personal computer systems. However, it is also true to recognize that no method can guarantee that an IS/NS is 100% immune to attacks of any kind.

Business Management and Leadership in the Digital Age[7]

Every organization, from Apple to Google, to the US Government, demands different and personal qualities in its management and

---

[7] Dr. Rene Naert, Professor at Alliant International University, contributed to this Section.

leadership team. In this trend, AI offers the exceptional capability to integrate conceptual and numerical LMI into an extraordinarily smart analytical powerhouse. AI applications act at both ends of the data management space, i.e., in the data acquisition and the subsequent data analysis phases. AI is expected to further increase the volume of data and the speed and accuracy for assessing that data. Digital technologies, such as process mining, digital twin, and robotic process automation will assist in making faster and smarter decisions. This can be accomplished by analyzing larger amounts of data, by more accurately identifying patterns and trends, and by providing deeper insights and more actionable recommendations. AI enables the automation of large complex tasks, the automation of schedules, and a more efficient utilization of resources, while concurrently evaluating risks and opportunities. All combined, AI maximizes and constantly monitors a system's performance and its progress. [15]

Leaders can now access enormous volumes of data, evaluate it rapidly, and derive insightful information thanks to AI-powered algorithms and corresponding fast analytics. This capacity enables leaders to respond fast and decide more accurately based on facts.

Knowledge creates power and awareness and strengthens people's emotional stability. The latter is a critical component of effective leadership. Leaders who are self-aware, empathetic, emotionally stable, and socially skilled are better equipped to build strong relationships within their teams and make sound decisions. [15]

Effective leadership involves a balanced combination of diverse traits and qualities that enable individuals to guide, inspire, and influence others toward achieving common goals. Each of these traits will benefit greatly by the enlightened usage of AI tools and processes. [1]

Leadership is a continuous journey of personal and professional growth. The latter has become more pronounced as digital technologies provide exceptional growth challenges and opportunities for becoming more effective leaders over time. [36]

With the massive integration of AI in the realm of Leadership practices, executives will Improve their efficiency and productivity outcomes by more effectively incorporating the AI potential into their executive skills-based arsenal. As AI becomes more common within the workplace, the efficiency of the application of organizational resources business operations will improve.

How are Leaders Using Artificial Intelligence Tools?

AI can help make faster and smarter decisions by analyzing large amounts of data, identifying patterns and trends, and providing insights and recommendations. One can use AI to automate repetitive tasks, optimize schedules and resources, evaluate risks and opportunities, and monitor performance and progress. [9]

How Can Artificial Intelligence Propel Management and Leadership Concepts/?

AI is poised to have a significant impact on management and leadership across various domains, including internal and external operations. As globalization creates globally distributed networks, AI will help to deal with complex multi-faceted risk issues, effective global collaboration, and improve worldwide customer experience. [7]

In the future, AI could potentially handle more strategic decision-making processes, such as analyzing market trends, identifying growth opportunities, and even making acquisitions. However, it's important to note that while AI may eventually replace some CEO responsibilities, it's unlikely to completely replace them. [36]

AI-based platforms are designed to empower decision-makers with predictive analytics, enable them to anticipate future outcomes and make proactive decisions. By leveraging historical data combined with machine learning algorithms, management can predict employee behavior and customer response, opportunistic market trends, and associated risks.

In summary, AI is set to transform management and leadership methods and procedures by providing leadership teams with exceptionally powerful tools to enhance their decision-making processes by affecting the full range of necessary tasks, such as communication, talent acquisition and management, and the Synergetic Five-Step Innovation Process. However, effective management and leadership must remain critical in guiding the ethical and responsible use of AI, foster creativity, and ensure that AI technologies serve the best interest of organizations and the society as a whole. [7]

Chapter 7
The Digital Age and Nano-Technology

Introduction

Throughout history, advances in knowledge and technology have always been the sources of new ideas for improving work processes as well as products and services. Technology has been the source of new knowledge, and advances in technology have been the result of the human urge to explore the unknown. Knowledge, technology, human intellect, and ingenuity form an integral system, where each affects each other, and they all grow together.

One may say that technological expertise expands in two directions. The first direction involves humans wanting to expand technological expertise to explore and understand the macro-cosmos and the micro-cosmos. The second direction is focused on technology applications. The latter has accomplished extraordinary progress in the healthcare arena. For example, micro instruments and computer aided imaging have enabled heart operations without opening a patient's chest. Heart operations have become outpatient surgeries.

In material science, technology has advanced to nano-technology. Nanomaterials are defined as matters that have at least one of its three physical dimensions fall into the range of 1 to 100 nano-meter (nm), where one nano-meter is 1 billionth of a standard meter.[8] For reference purposes, a one-nm dimension is close to the dimensions of molecules and atoms.

Nanometer-sized kernels of material have different properties than the properties of natural materials that we typically know of. Further, if nanometer-sized kernels of one material are intermixed with nanometer-sized kernels of another material, we can create now properties that do not exist in nature. Note that the mixture of nanometer-sized kernels of different materials resembles a physical mixture of materials; there is no chemical reaction involved. Traditionally, in order to achieve desired

---

[8] One nano-meter is equivalent to one billionth of a meter or 10^{-9} meter

material properties, different materials are combined by way of chemical reactions to form alloys. In contrast, nanomaterial technology employs physical mixtures. Nanomaterial technology is at the beginning of a very new science, and the opportunities seem extraordinary, although not fully understood yet.

We will be able to create materials with unparalleled ideal properties. Nanomaterials can be produced with outstanding magnetic, electrical, optical, mechanical, thermal, and catalytic properties that are substantially different from those of traditional materials. These nanomaterial properties can be tuned as desired by precisely controlling the size, shape, synthesis conditions, and appropriate functionalization.

Nanomaterial research has been pursued aggressively on many fronts during the past decade. The findings are confirming the expectations. Nanotechnology offers unforeseeable innovations in many areas of mankind's urgent needs. Unlimited opportunities may open up for the energy industry and accelerate the transition to carbon-free footprints for major industry segments. For example, nanomaterial-based crystals can be designed for extremely efficient photovoltaic (PV) conversion of sunlight into electricity. Nanomaterial-based PV cells, such as the emerging 3-D nano-pillar PV technology is expected to beat planar Silicon-based PV efficiencies that seem to be limited to the 20% efficiency range.

Nanomaterials can advance storage efficiencies for electricity by raising storage capacity and reducing weight of batteries. High-efficiency battery technology is critically important for augmenting a solar-based energy economy. That technology can drastically extend the range of electric automobiles and reduce electric automobiles' weight.

A major concern is the cost of manufacturing devices. Cost reduction options are being aggressively explored because the nanomaterial-related advances are not simply desired but are urgently needed to combat devastating climate change.

Nano-related visions have led to successes in advanced designs and manufacture of integrated circuits and in advanced electronic designs and manufacture of nano-time increments to enable multi-million concurrent voice communications via smart phones.

Nanomaterials

What are nanomaterials all about? Why are the material properties so hugely different from those of traditional bulk materials?

Research has disclosed that material properties depend on the interplay of surface and volume-based characteristics. The nanomaterials' exceptionally high relative surface-to-volume ratios play a dominating role for creating the unique material properties of nanomaterials.

With the formulas for volume (V) and surface (S) of a sphere with the radius r, we obtain for the surface-to-volume ratio:

$$V = 4/3 * \pi * r^3$$
$$S = 4 * \pi * r^2$$
$$S/V = 3/r$$

The result resembles the familiar One-Over-X correlation, i.e., as the radius r assumes smaller values, the surface-to-volume ratio will amount to high values and, eventually, for infinitesimally small radii, the S/V ratio approaches infinity. At those extreme conditions, the surface-dependent characteristics dominate the nanomaterial properties.

The three-cube chart in Figure 7.1 provides a visual illustration for the case when one takes a one-cm cube with a surface area of 6 cm² and considers the total surface of all the nanometer-sized cubes (60 million cm² = 6,000 m²) that would fit into the one-cm cube.

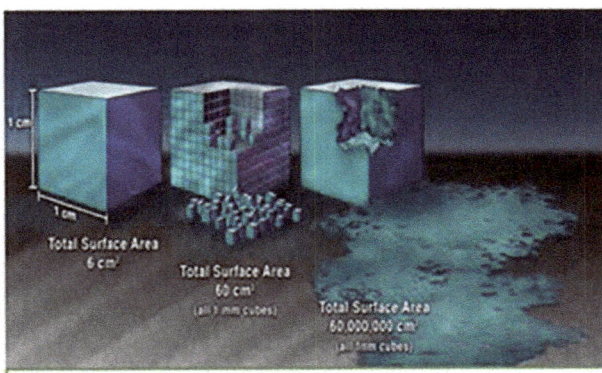

Figure 7.1   Nano-Sized Material Kernels   (Internet)

Translated into British Imperial units, the 6,000 m² would amount to 66,000 ft², which is equivalent to 23 full-size tennis courts and 85% of a soccer field. The increase in surface area created by nanometer-sized cubes inside a 1-cm cube is just enormous.

Let's consider what this means in terms of change for the surface-to-volume ratio.

The surface-to-volume ratio for the one-cm cube amounts to:

$$S/V_1 \quad = \quad 6 \text{ cm}^2 / 1 \text{ cm}^3 \quad = \quad 6 / 1 \text{ cm}$$

The surface-to-volume ratio of a one-nanometer cube amounts to:

$$S/V_2 \quad = \quad 6 \text{ nm}^2 / 1 \text{ nm}^3 \quad = 6 / 1 \text{ nm}$$

The ratio of the two S/V-ratios would amount to:

$$S/V_2 / S/V_1 \quad = \quad 0.01 \text{ m} / 1 \text{ nm} \quad = 1 / (10^{-9} / 10^{-2})$$

$$= 10 \text{ million}$$

The surface-to-volume ratio of a 1-nm-cube is 10-million-times larger than that of the initial 1-cm-cube.

The opportunities unleashed by nanotechnology are extraordinary and are being explored aggressively by research labs and pioneering companies around the globe. Two examples shall be discussed.

Concurrent Multi-User Mass Telecommunications in the Digital Age

The telecommunication industry's development and successful implementation of concurrent telecommunications are perfect examples of the Synergetic Five-Step Innovation Process, involving nanotechnology. At the very beginning, a vision toward society's urge for mass mobility and for mass communication on the move was recognized, technology's limitations were realized, an ideal technology-driven solution was envisioned, the solution was confirmed by logical-mathematical intelligence, and the technology development could commence.

Industry leaders like Qualcomm and others addressed the technological issue and mastered the challenge. Time Division and Code Division Multiple Access (TDMA, CDMA) are nanotechnologies that were initially developed in the 1980s and '90s to enable multimillion concurrent wireless telecommunications via the fairly limited number of available frequencies.

Humans extend their voice over relatively long time increments so that other people can recognize their voice and the words they speak. The thought experiment envisioned capturing a person's vocal expressions within nanosecond increments and to developing the electronics to break a person's voice into tiny increments. Each increment would be associated with a code so that the series of voice increments belonging to different people could be reassembled into normal speech periods at the receiving end. The short time increments are in the 400-nanosecond (ns) range, so that a single second entails 2.5 million increments. For a 4-second speech record there are 10 million time increments available. The result is that a large number of different communications can be transmitted concurrently.

High-Performance Computer Chips

Computer technology has become the ultimate enabler of thought experiment assessments by applications of LMI. Big-data analysis, process mining, and digital twin development apply AI to real-time process monitoring and control. This requires large data storage and high-speed processing capacities, which depend on integrated circuit chip capabilities.

A key factor for integrated circuit design and chip performance capability is the density of transistors, which translates into the size of transistors and their spacing. This is where nanotechnology has made a major impact on improving chip design and performance. Transistor sizes in the 35-nm range have been practiced for a while. The Taiwanese company, TSMC, a world leader in semiconductor manufacturing, has driven the transistor size down to 7 and 5 nm. The number of transistors that can be embedded into a quarter-inch-size chip reaches into the billions.

109

Chapter 8
Catastrophic Failures

Introduction

New technologies are primarily adopted because of significant economic advantages, but they also come with new features that need to be understood and must be harnessed to prevent catastrophic failures. The transition from one technology to the next typically involves transitions in management as well. The excitement about a new technology and its exceptional opportunities sometimes blinds management personnel sometimes to undesired yet silently present side effects. These side effects can be underlying natural laws, which give them an undeniable natural strength. Once man-made situations allow the side effects to be unleashed, humans cannot stop the resulting events.

Sometimes, blessings in disguise, which follow their own natural laws, prevent adverse side effects from unfolding and allow the desired outcome to occur. For example, Columbus' westward journey to India was heading straight into disaster, but a blessing in disguise saved the adventure, even though it created a false illusion of success.

The striking examples of four cases where no blessings in disguise could avert the catastrophic course of events shall be discussed in detail to illustrate the powerful natural force of neglected phenomena. Two of these examples involve technological transitions in the merchant marine industry and in the energy arena. Two further examples involve applied economics and address an inherent adverse economic phenomenon that was allowed to happen twice.

The Titanic's Multi-Faceted Failure Mode

Early in the 20th century, the merchant marine industry experienced a transition from sailing vessels to mechanically propelled ocean liners.

When steel, a material with superior structural strength, became available in the late 19th and early 20th centuries, the British shipping company White Star Line recognized the opportunity for ships with specially designed features that could prevent sinking. In response to their rival in the North Atlantic commercial trade, the Cunard Line's White Star Line developed the thought experiment to set a historic mark and create unsinkable ships. [33] White Star Line did not want to compete for speed. Their commercial ads focused on size, luxury and unsinkability, and they planned for three Olympic class luxury liners. The Olympic was commissioned in June 1911, the Titanic was commissioned in April 1912, and the Britannic's commissioning was planned for 1914.

The Olympic experienced several technically and economically successful voyages by the time the Titanic was launched. This success strengthened White Star Line's management in their conviction that their three-part philosophy was correct: large, luxurious, and unsinkable. The Titanic, the largest of the three sister ships, had the shortest life. It sank just three days into its maiden voyage from Southampton to New York.

The fate of the Titanic in April 1912 was very unique. Several warnings prior to departure and during the journey were

Figure 8.1   The Unsinkable Titanic on Sea Trials, 10 Days Before Her Departure Across the Atlantic   (Internet)

ignored. The early success with the Olympic played a role in those decisions. The concept of unsinkability had blocked management's mind. [33]

The unfortunate events leading to the final fatale strike against the Titanic must be seen in light of the marine industry's transition from the old tiller-based vessel control mechanism to the more logical rudder-based system.

During the first half of the 20th century, commercial operation of sailing vessels continued concurrently to the introduction of mechanically propelled vessels. The steering method on sailing vessels was accomplished by tillers. When the ship was expected to turn in one direction, the tiller had to be moved in the other direction. Steering commands for the helmsmen were given

Figure 8.2 Large Commercial Sailing Vessel (Internet)

as "tiller commands." On large sailing vessels, the tiller technology was maintained. Chains and winches were employed to accomplish the difficult physical task of controlling the vessel's massive tiller. The tiller command continued being practiced even when the tiller mechanism was replaced by a huge steering wheel. That meant that the steering wheel had to be turned in the opposite direction of the ship's intended direction. In some vessels, the steering mechanism was reversed, and a new more logical "rudder command" was introduced. The rudder command considered the fact that, in order for the ship to turn in one direction, the rudder had to be turned in the same direction. New steering wheel design followed that same principle so that the steering wheel had to be turned in the direction of the ship's intended direction.

Because of the two opposing steering design concepts and the two opposing steering commands co-existing in the industry at the time, helmsmen were faced with four confusing options of how to respond to a steering command until the rudder design and rudder command were established as binding international standards in 1930.

This confusion in the industry at the time had indeed an impact on the Titanic's fate. The granddaughter of the only surviving officer disclosed a dear family secret in 2013. As Louise Patten reports, according to her grandfather, the helmsman on duty at the time of the iceberg sighting interpreted the lookout's iceberg warning and his steering request the wrong way. [37]

1 - Confusing Industry Practices for Steering Commands Prior to 1930

The helmsmen for the Titanic's modern rudder command system came with all kinds of different experiences. Some had been trained on the old tiller command, and some came with experience with the modern rudder command. Even more confusing was the circumstance that on some merchant vessels that were already equipped with the logical rudder-based control mechanism, the commands were still given in the form of tiller commands. As a result, the helmsmen at the time of the Titanic had to fully understand their vessels' control mechanism, and there had to be a firm agreement on the steering command so that the helmsmen could know how the steering command had to be interpreted and executed. The question arises if White Star Line had a disciplined training program in place for the hired crew of helmsmen.

2 – The Naval Architect's Design Approach for Unsinkability

The Titanic and its sister ships were designed with the idea to create unsinkable ocean liners. The Titanic was equipped with a double-hull design, and the ship was divided into sections separated by watertight bulkheads. The designers figured that four sections could be flooded without causing the ship to sink. Further, the ship's double bottom was expected to protect it against damage in shallow waterways. With these extra precautions, the designers characterized the Titanic as unsinkable, and the White Star Line's management team strongly emphasized their luxury liner's unsinkability in their advertising stories.

The concept of being unsinkable became a paradigm and mesmerized the Titanic's design and management teams to the point that they made wrong decisions. For example, the naval architect, Thomas Andrews, suggested that the bulkheads were not high enough and should be extended to a higher deck. He also suggested that the number of lifeboats should be raised from 20 to 48. Both proposals were rejected by management because the Titanic was considered unsinkable as it was. Specifically, it was mentioned that the additional 28 lifeboats were not needed because the Titanic itself was considered the lifeboat.

3 - Coal Stored in Closed Bunkers Can Self-Ignite

At the Titanic's time, it was well known in the marine industry that coal stored in enclosed bunkers can self-ignite within about three weeks. Coal dust begins to oxidize and develop heat. When the coal pile is sufficiently large, the heat cannot escape from its central regions of the pile and heats up the entire pile. Given enough time, approximately three weeks, the entire pile of coal can engulf in flames. The coal's three-week self-ignition risk was not a problem for a one-week journey from Southampton to New York. However, the Titanic's speed trials and a strike in the local shipping industry complicated the schedule planning for the Titanic's maiden voyage. [30]

The sinking of the "unsinkable" Titanic was investigated by the courts in London and New York. The final outcome left several critical questions unanswered and resulted in a number of hypotheses about what may have really caused the gigantic vessel to sink.

Journalist Senan Molony was inspired by the Titanic's fate and uncovered the most authentic information about the great tragedy. He spent more than 30 years researching the sinking of the Titanic. [33]

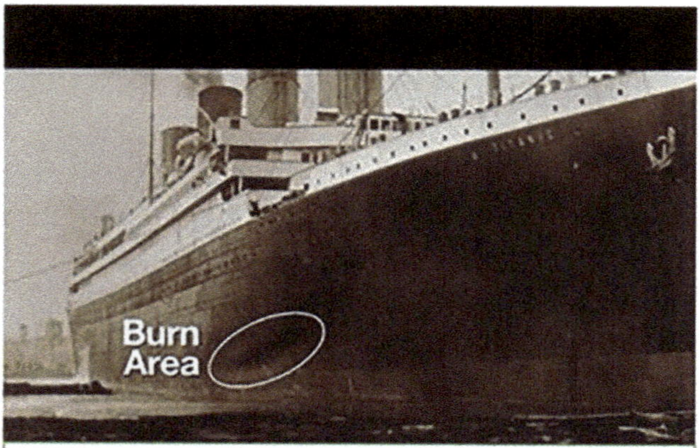

Figure 8.3 Titanic's Hull Damaged by Coal Fire Prior to Departure. (S. Molony [33])

He accidentally found photographs taken by the ship's chief electrical engineers before the Titanic ever left the Belfast shipyard. Molony was able to identify 30-foot-long black marks along the front right-hand side of the hull, just where the ship's lining was pierced later by the iceberg (Figure 8.3). Molony's profound 30-year research was able to trace those marks back to a coal fire in one of the Titanic's large coalbunkers.

Molony found that survivors who worked on the ship's engines cited a coal fire as the cause of the shipwreck. [33]   An official inquiry by British officials in 1912 mentioned it, too, but the narrative was downplayed by the judge who oversaw the case. Molony's research disclosed that he was a shipping interest judge. When he presided at a toast at the Shipwrights' Guild four years earlier, he said, "May nothing ever adversely affect the great carrying power of this wonderful country." Molony found in the court records that the judge closed down efforts to pursue the fire and made the finding that the iceberg alone was to blame.

According to Molony, "The official Titanic inquiry branded the sinking as an act of God." But he strongly felt that the Titanic's fate was not a simple story of colliding with an iceberg, so he continued his research. He visited marine institutes in England and Germany and reviewed data on shipping and weather-related data, and he talked with various professionals to obtain their input on weather conditions, material strength, coal fires, and the temperatures involved. [30] Metallurgical intelligence indicates that high temperature caused by coal fires against steel had a metallurgical effect on the steel, causing it to become brittle and reducing its strength by up to 75 percent.

Molony reports that in 2008, Ray Boston, an expert with more than 20 years of research into the Titanic's journey, said that he believed the coal fire began during speed trials as much as 10 days prior to the ship leaving Southampton. Molony also claims the ship was reversed into its berth in Southampton to prevent passengers from seeing damage made to the side of the ship by the ongoing fire. This made Molony believe that the fire was known to White Star Line's management, but it was downplayed.

Using Robert Boston's observations and the data obtained by Robert Ballard after he found the Titanic's wreck at the bottom of the ocean [3], Molony gained evidence that the iceberg struck the Titanic's hull in exactly the area where the weakness or damage to the hull had occurred before the Titanic had even left Belfast.

Molony concluded that Titanic experienced a perfect storm of extraordinary factors coming together: fire, ice, and managerial negligence.

4 – Blasting Forward at Maximum Speed Through an Ice Field

On its third day into its maiden voyage to New York, during the night from the 13th to the 14th day of April in 1912, the Titanic came to an ice field with occasional icebergs. A man was placed into a basket high up on one of the masts to look out for them. Visibility was impaired because of fog formation created by the temperature difference between the relatively warm air and the unusually cold water.

When ships enter an ice field, the typical reaction is to slow down the vessel in order to gain more time for course correction when an iceberg is sighted. The Titanic was not slowed down because of the threat of a coal fire and the need to reach New York as soon as possible. Consequently, the poor coal loading management from three weeks prior contributed to another poor decision. The Titanic continued blasting ahead through the ice field at its maximum speed of 22.5 knots. [33]

5 – Steering Blunder

When an iceberg was sighted, the iceberg report sent to the helmsman was combined with a definite steering command. Unfortunately, due to the many different steering mechanisms and steering commands existing in the industry at the time, the steering command was critically misinterpreted by the helmsman. Instead of steering away from the iceberg, the Titanic was now heading directly toward the iceberg.

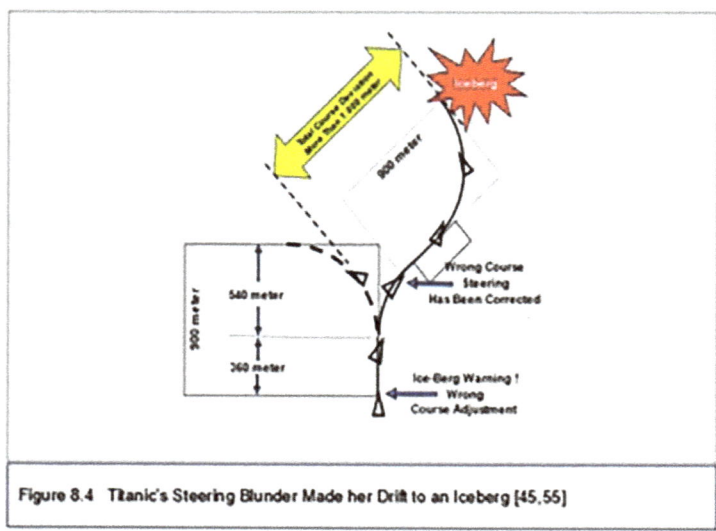

Figure 8.4 Titanic's Steering Blunder Made her Drift to an Iceberg [45,55]

A large vessel responds to course corrections very slowly, and when it does respond, it is very time consuming to reverse the vessel's direction. In addition, at full speed, large vessels have an extra-wide turning circle. Figure 8.4 illustrates what happens. [55]

The Titanic's first response to the false course correction was that the forward motion continued while the vessel rotated slightly into a drifting position, where the bow pointed into the new direction. This action required approximately 45 seconds while the vessel moved straight forward by approximately 360 meters. In the drifting forward motion, the vessel was pushed into the new direction.

Once the new direction of the vessel was clearly recognized, approximately 1.5 minutes had passed. During that time, the vessel covered approximately 540 meters in the forward direction for a total of 900 meters.

It took the Titanic 1.5 minutes to correct the mistake and another 1.5 minutes to attain the originally intended new direction. Considering the large turning circle of the large vessel at full speed, after approximately three minutes, the Titanic was more than a full km, i.e., 2/3 of a mile, closer to the iceberg than intended. The diagram in Figure 8.4 indicates that the Titanic was still in its left-turning mode, i.e., in a drifting mode to the right, when it touched the iceberg. Consequently, the Titanic did not just touch the iceberg, rather, it drifted into it with the full momentum of a large vessel moving at full speed.

The diagram in Figure 8.4 indicates how safe the Titanic could have been if the steering command had been interpreted correctly. The question remains whether Titanic could have prevented its sinking while being threatened by another looming disaster.

6 - Vessel Damage and False Decisions Made Under Pressure

The Titanic suffered severe hull damage on the starboard side. The cast iron rivets along 6 vessel segments were sheared off, allowing water to enter the six segments. Note that the ship was designed to remain unsinkable with four segments being flooded, not with six. The ship would sink so far into the water that the bulkheads were not high enough to prevent the water from spilling into the other non-damaged segments, too.

Considering the unsinkable design philosophy and fire in the coalbunkers, the Titanic's captain and chief vessel designer decided to maintain full speed. However, the bow wave created at full speed accelerated the flooding of the vessel. It has been said that the sinking could have been delayed until the arrival of another vessel if the Titanic had slowed down or stopped completely.

7 - Poor Visibility Confused Another Nearby Vessel

The poor visibility experienced by the Titanic at the time was the result of extremely cold water under relatively warm air. Random wavering or flickering of objects when seen through turbulent air is caused by atmospheric refraction, i.e., light is reflected differently by air at different temperatures. Since the local condition caused by turbulent air streams changes constantly, any object viewed in the distance can look completely different than what they really are.

There was indeed a freighter near the Titanic's site. Although the freighter received the Titanic's SOS signals, unfortunately, the freighter's personnel's view of the Titanic falsely indicated that it was not a passenger ship but a freighter. Consequently, the freighter's personnel decided to move on. [33]

8 - The Helmsman's Enormous Psychological Pressure

The fire in the Titanic's coalbunkers was discussed in court, but it was not considered a major circumstance affecting the Titanic's fate. The iceberg became the scapegoat for the legal matters involved.

As a learning experience for young entrepreneurs, it should be emphasized that the coal fires led to the full speed forward and for maintaining full speed even when the Titanic entered the ice field. And who would know if the helmsman on duty did not feel any fear for his life and acted under extreme pressure? He must have witnessed the frequent and urgent messages from the boiler room warning the ship's command about the enormous fires down in the bottom of the ship. Could his

118

pressure and fear have contributed to his false interpretation of the steering command? Or perhaps the lookout gave a false command in the first place?

9 - Severe Vessel Damage Caused by Extreme Coal Fires

The severity of the coal fires became obvious many years later, when Robert Ballard searched for the shipwreck on the ocean's bottom. Ballard's team found one of the boilers approximately one mile away from the Titanic's wreck, and a track of coal led them to the wreck. It appears that the Titanic was already severely damaged when it hit the iceberg. [30] If she had not hit the iceberg, could she have made it for another three days to the New York harbor?

Concluding Assessment of the Titanic's Failure

The negligent design decisions regarding the height of the bulkheads and the less than sufficient number of lifeboats contributed to the fatal consequences of the accident, but they did not cause the accident. The poor visibility and the flickering air did not cause the accident either, nor did the iceberg. And to be fair, the helmsman's wrong decision was not the original cause either.

There are two incidents that caused the string of unfortunate events that led to the iceberg affair. One tragic incident is the coal loading management. The coal was loaded several weeks too early. Once the coal had been loaded, no other reason for further delaying Titanic's departure should have been allowed.

The second incident is not a physical one but a mental position that influenced many decisions made during the vessel's design and construction, prior to departure and during the voyage. The concept of having an unsinkable ship caused unrealistic expectations and decisions. This is a very important point because similarly mesmerized mental positions cause accidents again and again in all industries and in political and economic policymaking environments.

Referring back to the Synergetic Five-Step Innovation Process, White Star Line and their chief naval architect had developed an ideal conflict-free solution as answer to the fierce competition in the transatlantic trade. While its competitors were competing on the basis of speed, White Star Line wanted to offer high-priced ocean crossings on unsinkable luxury ocean liners. It seems that the problem was that their LMI for assessing their thought experiment was not complete. The obsession with having an unsinkable ocean liner led them to ignore a disciplined coal management strategy. In contrast, they should have adopted the desire to strengthen the vessel's unsinkability by paying strict attention to a safe procedure everywhere, not just in their coal management strategy, but also in their helmsman training procedures. Considering all the facts, the coal management issue remains as the main cause of the fatal events.

There were plenty of warnings, but the obsession created by the notion of unsinkability prevented management's minds from being open for reconsiderations. Unfortunately, this is a behavioral phenomenon that occurs often.

The Titanic's chief naval architect, Thomas Andrews, was on board the Titanic with the intent of observing the vessel's performance and to identify improvements. When he learned that six compartments were being flooded, he must have thought then that his proposal of making the bulkheads higher could have saved the ship, his life, and that of everyone else on board. Survivors reported seeing him during the last moments and noticed that he was in extreme shock. One cannot help but being horrified by thinking about a chief designer, who had recognized the insufficient design but was not able to get it corrected and who was now experiencing the devastating failure of his biggest engineering creation.

An ideal conflict-free solution requires a full assessment using applicable logical-mathematical intelligence to prove the perceived solution and to install measures to actually achieve the solution's ideal and conflict-free characteristics. In today's digital age, AI offers exceptional opportunities to accomplish the task no matter how complex the task may be. AI can be employed to assemble complete product simulations, as well as toward safe and smart product designs. In addition, AI-enabled programs can be employed to simulate a number of accident scenarios to achieve true "unsinkability."

The Great Depression of 1929/30

Although macro-economics is a social science, explicit logical-mathematical intelligence underlies macro-economic phenomena. Examples of pivotal parameters are annual economic growth, inflation, unemployment, productivity growth, and the money supply. [8, 41] We shall discuss money supply phenomena and address the impact of money supply and its multiplier phenomenon on two economic derailments, i.e., the worldwide Great Depression in 1929/30 and the World Financial Turmoil in 2007/08.

The phenomena associated with the Great Depression and its causes have been analyzed and described elsewhere at great length. The purpose of revisiting the topic here is to illustrate the power of underlying Logical-Mathematical Intelligence and the possible impact of the Synergetic Five-Step Innovation Process on the outcome of a project. The underlying intelligence is of intangible nature. In contrast, the disastrous consequences are most visible and tend to cause great harm to individuals and to society as a whole.

To explore the stability of modern economies' financial systems, we need to look at the money supply and the banks' reserve rate. Banks use their deposits to assign loans to their clients. The reserve rate defines the portion of deposits that the banks should retain in their vaults to ensure that cash withdrawal requests can be met. The banks' experience indicates that reserve rates (RR) in the range of 2 to 5% are sufficient. To guarantee financial stability, the nations' central banks impose on the commercial banks a reserve requirement of approximately 10%. This rate varies between countries and between banks depending on the banks' size and their commercial business models.

Since the initial establishment of reserve rate laws and regulations in the early 1920s, money flow management is not just a cash flow issue anymore; money supply has gained an enormous impact on the entire economy's performance. Money supply is affected by a money multiplication phenomenon, which multiplies the money supply's impact on inflation, employment, the nation's Gross Domestic Product (GDP) and the economy's growth.

In this digital age, the money supply must be viewed through the eyes of its LMI. There are three variables to be considered. The first one is the actual amount of real money that has been deposited (G), and the second one is the money multiplier (MM). The third component is a form of virtual money (VM) that is created by banks using the debt creation model. Since the debt is supported by assets retained by the banks in return for the debt, the debt functions like deposits and must be included in the money multiplication formula. Thus, the effective money supply formula is written in the form:

Basic Money Supply (MS)	=	G + VM
Money Multiplier (MM)	=	1 / RR
Effective Money Supply (EMS)	=	(G + VM) / RR

The money supply formula indicates that the money supply can be affected and controlled by adjustments of the reserve rate (RR). Note that the MM formula is a One-over- X correlation, i.e., when RR approaches low values, MM rises exponentially and approaches infinity as RR approaches zero. Thus, the money supply is very sensitive to the reserve rate. The RR is indeed a very effective control mechanism that has long been installed and successfully practiced by nations' central banks prior to the world's Great Depression in 1929/30.

It is also important to note that the money supply depends foremost on the cash deposits G and the virtual money VM created by banks' collateral assets created by their debt programs.

Now, let's look at what happened in the 1920's leading up to the Great Depression, not just in the US, but in all European countries as well. There were several things that led to the breakdown of worldwide economies. We want to look specifically at the money supply as one of the culprits.

Workers in factories used to receive their wages in cash. In the '20s, banks were prepared to offer bank accounts to the working class, who then received certificates confirming that their earnings had been deposited in their banks and were available for withdrawal when they needed the funds. This shift ramped up the banks' deposits and gave them the opportunity to issue more loans. Their clients took advantage of the loans and purchased real estate properties and industry shares. Both were purchased under margin conditions, i.e., additional debt was created, which raised the deposits even more and raised the money supply accordingly. In essence, the economy got overheated. When the stock market experienced a decline, the margins created cash obligations for the clients. Since the cash was not available, they started pressuring the banks for more loans. When the reserves were utilized excessively, the money multiplier grew and contributed to the crisis. The vicious cycle had started, and the economy's instability took its course. The economic failure happened due to its own logic.

In summary, the money supply dynamics contributed greatly to the economic depressions around the world. Inflation rates skyrocketed into the 35% to 50% range and so did the unemployment rates. Looking at the political arena at the time, the economic disaster formed the perfect fertile ground for extreme autocracies across Europe.

The World Financial Turmoil of 2007/08

Although the world financial turmoil in 2007/08 was caused by a number of extreme liberations in complex financial systems, the money supply dynamics again played a major role in the collapse of the financial systems around the world. The money supply experienced essentially the same instability as in the Great Depression. The instability was caused by the same negligence of the same Logical-Mathematical Intelligence, except it was facilitated with a slightly different pretention. One may call it a "wolf in sheep's clothing."

The US government created a relaxed mortgage loan application and approval program. The "wolf's" name was Stated Income and Stated Assets (SISA). The initial objective was to make the mortgage loan application easier for people whose incomes and assets were difficult to document. Soon, banks applied the program to anyone who wanted to apply for a mortgage loan, including those who did not have the assets and income that they claimed. The program's nickname was "No Income and No Assets" (NINA). The program was applied on a large scale to the masses, and this is when the wolf lost his sheep's clothing. The virtual money (VM) portion of the banks' deposits skyrocketed and so did the economy's money supply. People over-leveraged their limited assets and headed into bankruptcies leading to the 2007/08 crisis.

In summary, the LMI was not considered, and the logic of the economic breakdown took over at nature's undisputable discipline. Although this was not the only mechanism that led to the financial turmoil, the result was that the money supply and money multiplication conditions of the Great Depression were allowed to happen a second time, albeit in a different coat.

The False Design of the Chernobyl Nuclear Power Plant.

The events leading to the nuclear accident of the Chernobyl Unit 4 in 1986 are surprisingly similar to the events leading to the Titanic's sinking. A critical review of the precursors of the Chernobyl accident indicates that the plant design and the misguided information communicated to the operations staff combined with a society's obedience to tradition created the conditions for an accident to happen.

The accident-prone pre-conditions of the Chernobyl story are described in great detail by Piers Paul Read.[39] Read's definitive account of the greatest environmental disaster of mankind provides the sequence of events that can be traced by this author's Synergetic Five-Step Innovation Process. The sequence of events resembles a suitable learning exercise for young entrepreneurs.

A brief introduction to nuclear physics and the uranium fission dynamics is required to illustrate the operation regime of safe nuclear energy generation and the dangerous regime that can lead to catastrophic failure. The threshold between safe and unsafe regimes demands utmost attention by program officials and plant designers to ensure that the reactor operation remains always within the

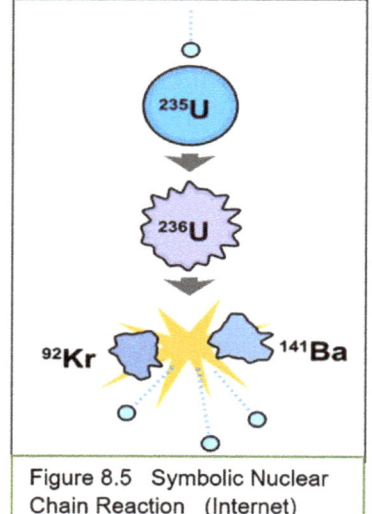

Figure 8.5 Symbolic Nuclear Chain Reaction (Internet)

boundaries of safe energy production. Fortunately, nature provides humans a wide window for designing safe power generation plants for peaceful utilization of the tremendous nuclear energy resource.

The chain reactions of uranium fissions are accomplished by neutrons. When a free-moving neutron penetrates into a uranium atom, it creates an instability and causes the uranium atom to split into smaller atoms as is illustrated in Figure 8.5. In this fission process, two to three neutrons are released, which are then available to cause two to three fissions in the next neutron generation. Since the neutrons are generated within approximately 10^{-4} seconds, one full second experiences 10,000 neutron generations, where the number of fissions can increase from neutron generation to neutron generation by the factor of 3 as shown in Figure 8.5. The result would be an astronomical growth rate for uranium fissions if one gave this process its uncontrollable free reign. There is no chance for humans to design a control system that can respond within a 10,000th of a second. Fortunately, nature offers mankind a beautiful control option, the said wide window for peaceful utilization of the nuclear energy resource.

While the majority of neutrons is released promptly within a 10,000th of a second by the fission process itself, a small portion of neutrons (approximately 0.65%) are released in a delayed fashion by the fission products. The delay periods range from a few seconds to several minutes, which results in an average delay of 12 seconds for all delayed neutrons combined. This process can be controlled comfortably by proven control systems and specially designed mechanical reactor control mechanisms.

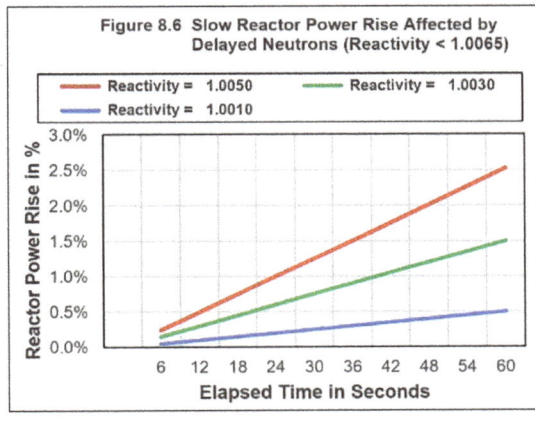

Figure 8.6 Slow Reactor Power Rise Affected by Delayed Neutrons (Reactivity < 1.0065)

The structural components of a reactor for peaceful power generation absorb neutrons so that not all neutrons (prompt and delayed) are available for causing the next fission. In addition, reactors are equipped with neutron-absorbing boron rods, which are used to effectively control the neutron population from one neutron generation to the next.

For reactor operation dynamics, the term "reactivity" was defined. Reactivity refers to the growth rate of neutrons from one generation to the next. A reactivity of more than unity indicates growth in the neutron population and the corresponding power of the reactor. A reactivity of less than unity indicates a decline in neutron population and the reactor's power. The boron-containing control rods are employed to effectively manipulate a reactor's reactivity.

For the control strategy of a nuclear reactor, the delayed neutrons, which amounts to 0.65% of the neutron population, plays a critical role. When the reactivity amounts to more than unity but less than 1.0065, the prompt neutrons provide a steady base of neutrons, but they alone do not affect the rise of the reactor's power.

The slow neutrons with a generation period of 12 seconds are in control of the reactor's dynamics. For example, if one allowed a safe reactivity of 1.0030, the delayed neutrons, being released at the average delay of 12 seconds, will cause a slow rise in power equivalent to 1.5% within 60 seconds, as shown in Figure 8.6. Thus, starting up a nuclear reactor in a safe mode is a very slow process. Shutting it down can be done faster, but there are other complications to be paid attention to.

In contrast, if the overall reactivity is allowed to amount to more than 1.0065, the prompt neutrons will yield a prompt reactivity that is larger than unity. Under that condition, the reactor's neutron generation period instantly drops from 12 seconds to one 10,000th of a second. The reactor dynamics become uncontrollable. For example, an overall reactivity of 1.0067 means that the reactivity of the prompt neutrons alone amounts to 1.0002 (i.e., 1.0067 minus 0.0065 = 1.0002). Under these conditions, the reactivity of the delayed neutrons is overwhelmingly surpassed by the prompt neutrons. As shown in Figure 8.7, when the prompt neutrons alone determine the reactor dynamics, they cause the power to rise by a factor of 7.4 (i.e., 740%) within one second. Depending on the actual amount of the prompt reactivity, the power could rise by a factor of 12 or more within a second. In such a scenario, the mechanical heat removal system and the monitoring systems cannot respond to the extremely fast neutron and power generation dynamics. In addition, the human response to confusing information can take a full second or even minutes. In the meantime, the reactor would be destroyed causing the release of radioactive materials into the environment and loss of lives.

Because of the exceptional risk proportions, an overruling reactor design principle has been defined and agreed upon, which specifies that the reactor design shall limit the reactor reactivity to less than 1.0065 for any possible operating situations, including any non-planned and accidental deviations from

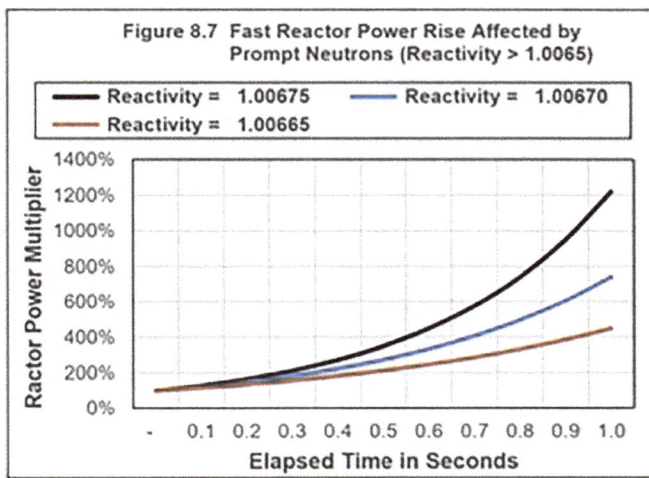

Figure 8.7 Fast Reactor Power Rise Affected by Prompt Neutrons (Reactivity > 1.0065)

normal operation. More than 400 commercial nuclear power plants have been in operation throughout the world since the 1960s and have proven that the knowledge of nuclear physics and corresponding nuclear engineering expertise have been able to design the reactors with the said reactivity limitation. No reactor had experienced prompt reactivity.

With that basic understanding of reactivity, we can now look at the events leading to the nuclear accident at Chernobyl Unit 4.

The Chernobyl reactor design was affected by political ambitions of the Soviet Union during the cold war period and a severely competitive mindset at the time. Nuclear program managers were not willing to sacrifice economics for safety. However, the public and operations staff were made to believe that the reactor was extremely safe, safer than the reactors of the capitalist West.

When the Chernobyl-type reactor design was proudly presented at an international conference held in Geneva in 1971, the reactor design sparked several critical concerns. Conference attendees expressed their concerns about:

- The size of the reactor and the possibility of locally different and shifting reactivity values.

- The possibility of a dangerous coolant-imposed reactor instability. It was feared that a rise in power could be accommodated by a coolant-imposed additional power rise, such that a self-feeding rise in power could be initiated.

- The lack of a containment building.

The comments made by international experts were either dismissed, overlooked, or ignored. In addition, unbeknown to the international experts reviewing the then incomplete information about the Chernobyl-type reactor, the reactor design did not exclude the possibility of reaching prompt criticality. The program directors did not want to accept the economic penalties associated with a reactor design, where the reactivity would be limited to the safe delayed-neutron reactivity. Further, the program directors did not share that unsafe aspect of their reactor design with the reactor operators. Instead, they proclaimed that their reactor was safer than the reactors of the capitalist world, despite the fact that the opposite was true. [39]

Because of the generously designed reactivity reserve, the possibility existed that, under adverse operating conditions, the reactor could experience prompt criticality. These dangers existed unbeknown to the operating staff. They thought that their reactor was indeed "absolutely safe."

The staff's safety perceptions were confirmed by several years of safe operations of the Chernobyl-type reactors in Chernobyl and elsewhere in the Soviet Union. The accident-prone conditions had not yet surfaced, until the moment when it was decided to run a risky experiment with the commercial-size Chernobyl Unit 4 to prove its "exceptional" safety under purposely imposed severe emergency conditions.

In general, the International nuclear energy community was and still is concerned about sufficient reactor cooling when the coolant flow is interrupted for some reason and the reactor is shut down. The reactor's after-heat is substantial and continued cooling after reactor shutdown is critically important to prevent reactor meltdown. The US Atomic Energy Commission at the time had conducted many tests with non-nuclear and nuclear test facilities, i.e., small test facilities, not with commercial-size plants, to simulate the after-heat phenomena and to design emergency cooling systems and corresponding safe operating procedures.

On the day of its accident in 1986, the Chernobyl team had to take the Unit 4 through a regular shut-down procedure for maintenance and refueling purposes. The idea was approved to use the occasion to perform a loss-of-coolant experiment during the shutdown phase. The idea of the emergency cooling experiment was to confirm that sufficient power could be squeezed from the main turbines while they were spinning to a halt to bridge the time gap until the emergency cooling system would kick in. The execution of this plan unleashed the plant's insufficient safety provisions. The reactor's own logic took over and instigated the plant's horrific failure.

One of a nuclear reactor's own logic is known as the Xenon poisoning effect. Xenon is a fission product that features a strong neutron absorption combined with unique dynamics of its presence in the reactor. Because of its strong neutron absorption and its unique concentration dynamics, Xenon affects the reactor's reactivity and, thus, is an important factor to be strictly accounted for in reactor operations.

During steady reactor operation, the Xenon presence reaches an equilibrium after approximately 40 to 50 hours.[61] When the operators want to increase power, they pull control rods somewhat to raise reactivity. In response, the Xenon reactivity decreases initially, which needs to be considered by the operators. When the operators want to reduce power and insert control rods to reduce reactivity, Xenon concentration increases and, thus, reduces the reactivity further. The final extent of Xenon's negative affect on reactivity depends on the rate of power reduction imposed by the control rod insertion. The Xenon poisoning effect for a fully shut down reactor is so strong that the limited reactivity reserve of less than 0.65% is not sufficient to override the Xenon poisoning so that the reactor cannot be restarted for approximately ten days. Thus, reducing the power level to an abnormally low level and keeping the power steady at that level is a very difficult task for the operators. First of all, in order to keep the Xenon poisoning under control, the power reduction from full power has to be done at a slow pace.

During power reduction at Chernobyl Unit 4, the reactor's power had suddenly dropped below the power level at which the experiment was to be executed. The strong power reduction is typically a normal and not a concerning effect during regular shutdown operation. However, this was an unwanted effect at Chernobyl because they wanted to hold the power at a low level. As the operators tried to raise the power again, the reactor did not respond, and the operators were clueless as to what was happening. The operators withdrew more control rods but because of the reactor's ongoing strong Xenon poisoning kicking in, the control rod movements remained ineffective. The continuing reduction in power could not be stopped. In desperation and because of his firm conviction that the reactor is extremely safe, the supervisor was determined to raise the reactor's power so that the test could be performed as planned and as committed within the authoritarian regime. He requested the operator to pull more control rods. Being aware of the possible consequences, the operator hesitated, but he was firmly ordered by his supervisor to pull the control rods. As the operator followed his orders, nothing seemed to happen at first. [39, page 64].

In a safe reactor, where the reactivity can never be raised beyond the delayed reactivity level, it is not possible to raise the power level once the Xenon poisoning has been built up. However, the Chernobyl Unit 4 reactor had more reactivity reserves and had the capacity to override the Xenon poisoning. As the control rods were extracted further to raise the reactivity, the effective reactivity started rising and caused the Xenon poison to reduce quickly. The reactor instrumentation suddenly indicated a strong rise in power, which was noted by the operators. However, under the ongoing conditions, the natural reactor dynamics were extremely fast. By the time the reactor instrumentation indicated to the operators that the reactor power was rising again and by the time the operators were able to issue a response for inserting the control rods to limit the rise in power, the reactor's own logic had already taken over and could not be stopped anymore. The dynamics shown in Figure 8.7 illustrate that, under prompt reactivity conditions, the reactor destroys itself within a fraction of a second. The operator's attempt to insert the control rods were unsuccessful because they were stuck. Concurrently, the water-based heat removal system was suddenly exposed to an extraordinary heat load and responded with its own dynamics and caused unstable feedback between various components in the heat removal system. "There was a thud, followed by further thuds from deep inside the reactor building. The walls shook, the lights went out." [39, page 65]

Had the supervisor not been so obsessed about the absolute safety of his reactor, he may not have given the strict order to pull so many control rods. The communicated perceived "absolute safety" created a mesmerized mindset and shielded the reactor's faulty design and the officials' irresponsible loss-of-coolant experiment to be conducted with a full-size commercial power reactor.

The similarity between Chernobyl and the Titanic is striking. For the Titanic, the communicated perceived "unsinkability" created a mesmerized mindset and shielded the vessel's faulty design and the management's poor decision-making during several critically important phases of the vessel's operation.

An ideal conflict-free solution requires a full assessment using applicable logical-mathematical intelligence to prove the perceived solution and to install measures to actually achieve the solution's ideal and conflict-free characteristics.

AI in the digital age offers exceptional opportunities to accomplish the task no matter how complex the task may be. AI can be employed to assemble complete product simulations, including complex nuclear reactor dynamics. In addition, AI can be employed to safeguard complex system operations, such as nuclear reactors. For example, advanced nuclear power reactor operations are constantly automatically "supervised" by AI-enabled programs to ensure that the reactor is in stable operations across all regions within the reactor and that operator-initiated controls are within acceptable ranges. Any deviations from the norm are instantly recognized, and an orderly reactor shutdown may be initiated by AI-enabled safety control programs.

Chapter 9
Blessings in Disguise

Introduction

Blessings in disguise are typically hidden deep down under the cover of successful innovations and may be uncovered when catastrophic disasters are suddenly prevented. The events, independent of good or bad, occupy people's minds and thinking so that the disguises, if any, remains unrecognized. It may take time for the fortunate disguise to enter people's awareness and be recognized as fortunate blessings after all.

A blessing can play into the progression of events and prevent the disaster. In that case, the main actor in the project collects all the credit without anybody knowing that he was actually heading into disaster. It takes detailed analysis and the collection of hidden facts and unique circumstance to eventually disclose the tremendous risk involved in the project. Columbus' westward sea passage to India is a prime example. If the American continent was not in his way, he would have sailed into nowhere.

A project may have several parties involved, each with different roles, although each having the same goal in mind. Each party may experience different consequences for each of the potentially different outcomes. A negative outcome of one aspect may represent failure for one party and success for the other and vice versa. It all depends on which aspect fails first because failure of one aspect may prevent failure of the other. The failure that is prevented by the first failure may never enter the mind of the parties nor the judges trying to identify the victims and the accused. Again, the sinking of the Titanic is a prime case.

Failures and blessings in disguise are project phenomena, which, because of their hidden nature, are often not recognized as potential outcomes. The awareness of hidden failures and potential blessings in disguise can lead to special analysis aimed at their discovery beforehand.

The Survival of Christopher Columbus

Evidence discovered in Newfoundland confirms that the Vikings discovered America well before Columbus, but it was Columbus' discovery that changed the world.

Columbus is undoubtedly credited for discovering America for the Europeans. And the story presented in Chapter 2, Historic Emergence of Innovations, is consistent with typical descriptions in honor of his pioneering accomplishments. His discovery that time and distance around the globe are interconnected and his pioneering move of adding new technology in the form of the recently invented mechanical clock to his navigational tool set are simply remarkable.

There are several inconsistencies in Columbus' behavior and his entrepreneurial decision-making. Inconsistencies in people's logic and their persistent neglect of contradicting information typically result in high risks for the adventure and can cause complete failure, but not for Columbus.

Columbus' extremely low estimate of the westward distance from Portugal to the Indies, his accurate determination of the actual distance travelled to the Haitian Islands, and his persistent claim to have reached the West Indies, all remain a mystery that deserves further investigation.

The LMI available during Columbus' time and especially available to Columbus himself, and the Synergistic Five-Step Innovation Process, shall be applied to trace the events involving Columbus' explorations and his discoveries with the objective to shed further light on the mysteries surrounding Columbus' momentous historic episodes.

During Columbus' time, tradesmen practiced cumbersome land-based travels to India to bring valuable spices to Central Europe. Overland trade routes led through Turkey and Iran. A sea passage was practiced leading along the African coast and around the Cape of Good Hope. Both travel routes were time consuming, cumbersome, and risky. The emerging perception that the Earth is not flat but round like a sphere gave Columbus an alternate idea for reaching the West Indies in a different and less risky way.

The thought experiment in Columbus' story Is sound. If the Earth is indeed round, there should be a westward sea passage from Europe to India. Thus, the first part of the Synergetic Five-Step Innovation Process appears to be fulfilled.

The first issue is seen in Columbus' assessment of the size of the Earth and his estimate of the westward distance to India. There were two estimates of the size of the Earth. The larger estimate was close to the actual size of the globe while a second estimate was about 25% smaller. Columbus defended the smaller estimate, and it is not quite clear if he really believed in the smaller estimate or if he just pretended to believe in it. One can think of two reasons for his misrepresentation of the Earth's size. He needed to raise financial support for exploring a westward sea passage to India, and he probably wanted to make the adventure look easier than it really was. Second, he may have understood very well that he had to advocate an easy and less risky journey in order to be able to hire a crew for the journey into the unknown. This behavior continued during his entire journey. He maintained two log books, one for his crew to see, where he consistently reported less miles traveled, and one for himself.

On the other hand, Columbus was one of the few people of his time, if not the only one, who had made the connection between a clock's time and the distance traveled in east-west directions around the globe. He understood that distance around the globe could be determined by the time differences between a particular location on the open ocean and the home port. This conclusion can be made because Columbus made his proposals for the exploration of a sea passage to India based on his new entrepreneurial knowledge. In addition, during his journey to the West Indies, he practiced exactly that new knowledge and, after arriving at the Haitian Islands, he recorded the actual miles travelled extremely accurately, at about 4,000 miles.

The trade peoples' eastward travels to India over land and over the ocean around Africa must have given Columbus some information about the eastward distance from Portugal to India. Further, the extent of the continent's size was estimated to cover 185 degrees of the globe leaving half of the full circle for the open ocean. Using the larger estimate of the Earth's size, which was quite accurate at the time, would indicate that the westward sea passage from the west side of the huge continental mass to its east side would amount to half the Earth circumference, i.e., approximately 12,000 miles. Using the 25% smaller estimate of the Earth's size would indicate a westward distance to the West Indies of 9,000 miles. Yet, Columbus promoted in his proposals to potential sponsors a 2,400-mile distance. There does not seem to be any records for the basis of that low estimate. The only justifications are seen in Columbus' intent to make the exploration look easier than it was. Columbus' actual travels compared to the full westward sea passage to India is illustrated in Figure 9.1.

It did not seem to concern Columbus that the Catholic monarchs, who he approached for funding for his exploration of a westward sea passage to India, thought that his 2,400-mile estimate was too short by a factor of four. According to the prevailing knowledge that the huge contiguous continent covered 185 degrees, the monarchs' estimate was reasonable. Why didn't Columbus revisit his own analysis and adjust his estimate?

Columbus reported that he travelled for approximately 4,000 miles when he made land fall. His logbook mentions the name of the San Salvador islands, but it is assumed today that he landed at the Haitian Islands, which are approximately 3,900 miles away from Lisbon, Portugal. Thus, Columbus' navigational skills and distance measuring technique involving the clock proved to be accurate.

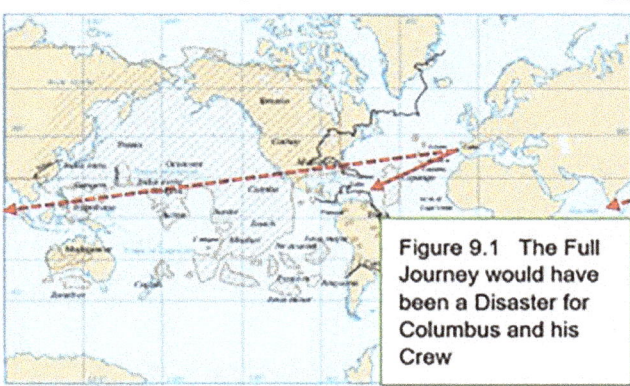

Figure 9.1 The Full Journey would have been a Disaster for Columbus and his Crew

It is interesting to note that the Haitian Islands can be reached by way of a southwesterly course from Portugal and that they are located at the same latitude as India. Did Columbus steer his ships on purpose along a southwesterly course hoping to reach India after approximately 4,000 miles? Or did that outcome simply happen by chance?

135

Columbus seemed to understand that the globe requires 24 hours per revolution and that a time difference equivalent to one hour translates into a distance equivalent to 24th of the globe's circumference. His correct determination of the distance traveled to the Haitian island indicates that he knew that one hour of time difference indicates 1,000 miles traveled.

If there was no Haiti and if there was no other island or continent in Columbus' way to India, he would have been sailing into nowhere and had no chance to return home. The fact is that the eastward journey from Portugal to India represents a distance equivalent to approximately 20% of the Earth's circumference. That means that the westward journey is equivalent to 80%, i.e., the westward journey amounts to approximately 19,000 miles. Columbus would have had to add 15,000 miles to his actual 4,000 miles. Was he possibly aware that the Vikings had been to the Americas many years before him? [26]

But Columbus was saved by a very fortunate blessing in disguise. Not only were a few islands in his way, but a complete continent was in his way, allowing him to declare, "mission accomplished."

The White Star Line's Survival after the Titanic's Sinking

The Titanic's fate is discussed in its full detail in Chapter 8, Catastrophic Failures. The series of events involved in the luxury ocean liner's fateful maiden voyage clearly illustrate what Dietrich Doerner describes as "The Logic of Failures." [10] Nine individual events, each one totally independent of the others, all combine into a chain of incidents that seem to be coordinated by an "invisible hand." The chain of uncontrollable incidents is eventually led to an unavoidable fulfillment of the chain's logic. [33]

Surprisingly, even though the Titanic went under and was not saved by a blessing in disguise, Michelle Bruneau's concept of "Blessings of Disaster" [6] was involved. While it did not save the Titanic, it saved its owner, the White Star Line.

When the Titanic's survivors were interrogated in court in Great Britain and in the US, the steering blunder was not on the table. The only surviving officer who knew about it did not reveal that information at the time of the court hearings. He wanted to protect White Star Line and wanted to prevent the company's bankruptcy. [37]

The coal fire was mentioned during the court hearings by surviving crew members, but the full extent of the coal fires and the damage they had already created by the time the Titanic drifted into the iceberg remained unknown. The coal fire and related vessel damages were not considered by the courts as a decisive circumstance.

Thus, the iceberg and the poor visibility at the time were considered the cause of the Titanic's sinking. As a result, the White Star Line company was able to survive financially.

Many decades later, the findings of Robert Ballard's deep-sea research revealed that the coal fire had already penetrated the vessel's double bottom, and coal had dropped out of the bottom for at least a full mile. Had the Titanic not hit an iceberg, she still may not have been able to reach New York Harbor. In that case, White Star Line's management may have been found guilty of irresponsible coal management procedures. In light of the likely outcome of the alternate scenario, where White Star Line may have been found guilty, the iceberg event offered a tragic but convenient scapegoat.

In this book's context of analyzing successes and failures of pioneering projects, the steering blunder and the iceberg incident could be considered a "blessing in disguise" for White Star Line, a blessing that allowed them to stay in business. If there was no iceberg, the fire in the coal bunkers would have caused the ship to sink and White Star Line would have been blamed for the disaster. If the iceberg had been avoided by a correct interpretation of the steering command, there would not have been any excuse for White Star Line. It is interesting to note that the fires in the coalbunkers were the reason for blasting through the ice field at full speed and possibly also for the steering blunder. Therefore, the coal fires had an impact on the vessel's management team and the helmsman's steering blunder and, thus, can be considered the decisive circumstance that created the blessing in disguise for the White Star Line organization.

Chernobyl and its Disregarded Blessings in Disguise

The events leading to the nuclear accident at Chernobyl's Unit 4 reactor entail several opportunities for positive turns away from disaster and toward a positive outcome. However, these opportunities were not recognized as such because the need for such opportunities was deeply denied. The USSR's nuclear program at the time was structured during the cold war period in isolation and in a spirit aimed at accomplishing more than the West.

The first great opportunity was offered in 1971 at the international conference in Geneva. The logic of failure inspired a USSR delegation to attend the conference where they presented their reactor design with pride of their accomplishments. The idea of benchmarking their design to identify improvements did not enter their mind. The critical comments about possible reactivity instabilities inherent to their reactor design and the lack of a reactor containment did not reach their consciousness. As it turned out, these comments addressed the primary cause of events that led to the accident and the subsequent uninhibited release of radiation.

Another opportunity for a positive turn of events came into being when the power dispatcher called and terminated the reactor shutdown just minutes prior to the execution of the fateful loss-of-coolant experiment with a full-size commercial reactor. Unfortunately, the test was postponed, not abandoned. The next shift was charged with the test execution, although that staff had not been prepared for the complexity of the test.

And finally, in the second approach to the test, when the reactor dynamics had shifted into the unknown and had dropped its power level much below the level at which the test should have been performed, the operator's first reaction to the supervisor's request for pulling the control rods out of the reactor was the right one: "I should not do that." This would have been the last silent blessing in disguise that could have saved the entire situation. But the fear that comes with the demand for absolute obedience caused the supervisor to strengthen his request and made the operator obey the order. At that moment, the accident more than 20 years in the making took its final course. The reactor reached prompt criticality and instantly destroyed itself and its environment.

Here is the LMI that, unbeknown to the misguided supervisor, determined the situation at the critical moment just prior to the accident.

Logical-mathematical intelligence involving reactor dynamics, including the dynamics of Xenon poisoning, could have warned program officials and operations staff about the complexity of the reactor dynamics during reactor shutdown and the added complexity created by the intent to perform an experiment that is extremely complex by itself. A reactor that is designed with a limited reactivity reserve of less than 0.65% would not have been able to override Chernobyl's strong Xenon poisoning. In addition, full knowledge and understanding of the Xenon poisoning dynamics should have told the reactor workers at Chernobyl that if the power had to be held constant at a low level, the reactor shutdown must be accomplished extremely slowly so that the Xenon poisoning could be maintained at a manageable level.

Artificial Intelligence and digital twin technology available in the digital age give scientists and nuclear reactor operators the perfect tools to exercise logical mathematical intelligence in parallel to actual facility operations.

Chapter 10
Project Execution

Introduction

This chapter offers a few choices for initial assessments of proposals for new products, new services, new processes, and completely new enterprises. There are a number of platforms on the market with wide ranging capabilities and an assortment of built-in functionalities. Since different industries have very different specific requirements, specialized industry platforms have been developed.

Independent of an industry's unique circumstance, for initial assessments of innovation proposals, one can use less complicated and easy-to-use analysis platforms. This recommendation applies to professionals in various industries and particularly to students enrolled in business management programs.

Since different analyses typically require different platforms, for initial scoping analyses of innovation proposals, one should consider a family of generic platforms that are compatible with each other, so that data can be transferred from one platform to another.

Microsoft offers a family of generic and easy-to-use business analysis platforms, which are compatible and allow transfer of data from one to the other. These include the well-known MS-WORD and MS-PPT applications and the generic analysis platforms MS-EXCEL and MS-PROJECT, which are briefly introduced below. For more project analyses, Microsoft offers expanded analytical capability and data storage platforms, which are not discussed here.

Practicing Logical-Mathematical Intelligence

Microsoft EXCEL is a powerful app to process LMI and assess the validity, benefits, and risks of perceived ideal and conflict-free solutions of a wide-range of industrial and societal issues. The platform features an endless grid of cells that are organized in numbered rows and letter-named columns. The MS-EXCEL platform is particularly easy to use for initial assessments of selected business strategies.

MS-EXCEL provides the possibility to specify arithmetic functions to manipulate data contained in the cells. One can define any function that represents one's LMI. In addition, MS-EXCEL contains the algorithms of approximately 500 functions that a user can call up and utilize to exercise polynomials and perform regression analyses, to name a few.

An important component of idea assessments is the determination of its business model's financial performance. Implementing new products and processes requires upfront investments, and potential investors always want to know if new business endeavors can earn enough profit so that investors can earn expected returns on their investments. MS-EXCEL is set up to determine the various financial performance projections for investors. The most important one is the return on investment (ROI), although, in MS-EXCEL, it's called internal rate of return (IRR). Further, the platform has the built-in capability to generate various forms of business performance charts, including the ROI, IRR Chart.

For example, the chart in Figure 10.1 illustrates the iteration underlying the ROI calculation. ROI cannot be determined in a single calculable step. It is determined by an iteration. If done manually, one chooses an arbitrary discount rate (DR) and calculates the projects net present value (NPV). Several such calculations can be

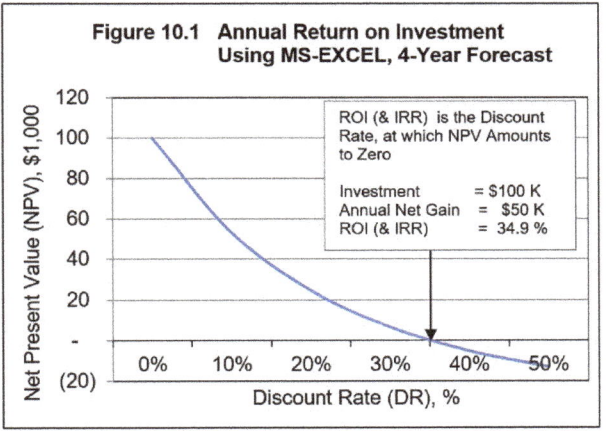

Figure 10.1   Annual Return on Investment Using MS-EXCEL, 4-Year Forecast

plotted as shown in Figure 10.1. The discount rate, at which the NPV turns zero, marks the ROI (or IRR).   Carrying out this iteration process manually is a time-consuming affair. MS-EXCEL offers a built-in process that executes the iteration in an automated fashion.

A single MS-EXCEL worksheet is called a spreadsheet, and a series of connected spreadsheets make a workbook. Algebraic correlations used in one spreadsheet may utilize data contained in another spreadsheet. One can dedicate a spreadsheet to one corporate department and other spreadsheets of the same workbook to other departments. Thus, the spreadsheets of a workbook can be used to simulate a complete business model and study its various performance objectives under different scenario strategies.

In the context of AI and its impact on decision-making, platforms like MS-EXCEL contain a number of logical decision correlations. Without any sophisticated programming skills, one can devise a series of nested logical decision processes and attain an intelligent business analysis model, which is suitable for conducting a quick initial analysis of a perceived ideal and conflict-free solution. Thus, MS-EXCEL can accept a fairly detailed and deep reaching LMI and offer an initial yet sound practice of AI. Although the AI applications in a tool like MS-EXCEL represent a simple and limited use of AI, it offers a convenient exposure to what AI essentially means and what it does.

The results can be charted within MS-EXCEL to generate typical performance curves. Best of all, the charts can remain connected to the correlations used for the business model. As the assumptions for the business model are changed and new results are obtained, the charts will automatically reflect the modified results. Charts can be organized as curves, pie-charts, or bar charts. The charts can be done in 2-D and 3-D formats. Charts can be transferred to MS-WORD as part of project reports and to MS-PPT for project presentations.

MS-EXCEL is a very diverse, powerful app that can be used to quickly scope a new idea or a new process. Since the results can be assembled for transparent visualizations, MS-EXCEL makes a great tool for creating powerful presentations to promote ideas. In fact, no idea should be presented without an initial analysis illustrating its value and its opportunities. Once an idea passes an initial MS-EXCEL-based low-cost assessment and proves promising, more costly sophisticated analyses are justified.

Project Management Principles and Standards

Project management (PM) is a broadly practiced art and science. [16,38] A number of PM platforms have been developed for different industries. The platforms are collections of tools that help to predict and control the outcome of projects of all sizes.

142

The Project Management Institute [38] provides a holistic and realistic approach to project management that combines the human aspects and the culture of an organization. It covers concepts and skills used to schedule project tasks, plan resources and budgets, and lead project teams to successful completion of projects. [29]

The project management profession has significantly evolved due to emerging technologies, new approaches, and rapid market changes. Reflecting this evolution, the standard for project management enumerates 12 principles of project management. The PMBOK® Guide is structured around eight project performance domains. [16, 27, 38]

While the specialized industry-oriented project management platforms, such as for example, Oracle's Primavera software, are comprehensive enterprise project management (EPPM) tools, Microsoft developed a generic platform, MS-PROJECT, which is very practical for a wide range of applications and has strong graphic display features.

PROJECT is a project management software designed to assist project managers in developing project plans, assigning resources to tasks, tracking progress, managing budgets, and analyzing workloads. It has features such as Gantt Charts, kanban boards, and project calendars for project management professionals.

Gantt Charts are EXCEL-type spreadsheets containing rows for tasks and columns of time increments. The individual tasks are marked by horizontal bars reflecting the length of time scheduled for the respective tasks. Each bar can be linked to other bars to reflect the dependency between tasks. If one marks the planned time duration of each task and makes the beginning of each task dependent on the completion of its pre-

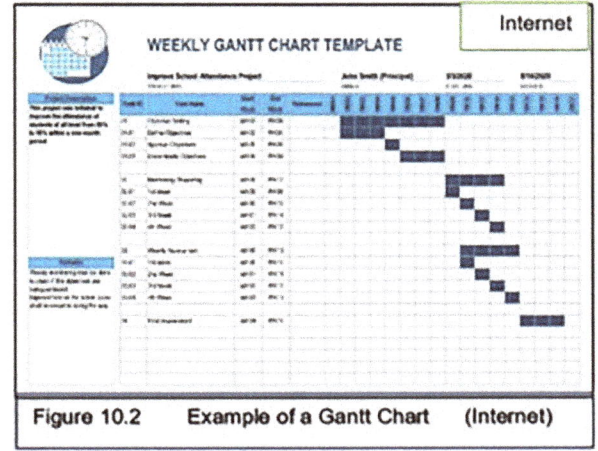

Figure 10.2   Example of a Gantt Chart   (Internet)

143

curser, one can change a task duration, and the entire project is adjusted accordingly in an automated fashion.

This software allows the determination of the critical path and, thus, facilitates an efficient measure to shorten the overall project schedule. This is an important project management task because short project cycles are an important aspect of an enterprise's return-on-investments and its overall financial performance.

The PROJECT platform includes an integrated cost/schedule system. It monitors three cost/schedule parameters: the budgeted cost of work scheduled (BCWS), the budgeted cost of work performed (BCWP), and the actual cost of work performed (ACWP). These parameters can be used to assess the project status by determining the cost variance (CV) and the schedule variance (SV). At the time mark of 50, the example in Figure 10.3 provides the 54% and 24% for cost and schedule variance, respectively.

Cost Variance (CV)   =   ACWP - BCWP   = 54%

Schedule Variance (SC)   =   BCWS - BCWP   = 24%

The ACWP tends to exceed the BCWS, and the BCWP tends to undercut the BCWS as is the case for the example in Figure 10.3. The variances happen because of the inherent nature of projects combined with the nature of human behavior. It is difficult to get a project started, and things just do not work smoothly at the very beginning. This inherent defect of project progression causes humans to correct the mistakes by throwing money at them in the hopes of straightening things out. Unfortunately, that behavior accomplishes just the opposite and adds to the overall variance of the project.

The cost/schedule graph in Figure 10.3 clearly illustrates what the project manager's job really is. He needs to design the BCWS curve such that it starts at a low slope as shown in Figure 10.3 and that he designs a steep slope in the middle of the projects time allotment to catch up and be able to finish the project in the shortest possible time. This is where the integration between Gantt Chart design and cost/schedule planning mark the project design phase.

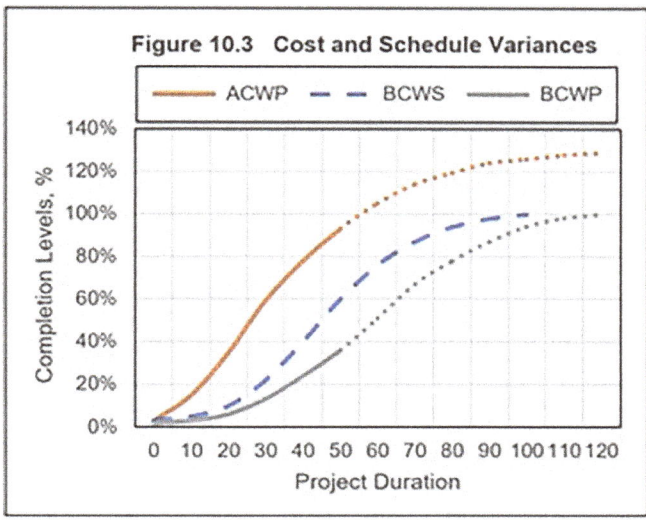

Figure 10.3   Cost and Schedule Variances

When a project crosses the starting line, the project manager has to be omnipresent and make sure that every task is completed as intended, i.e., on schedule and within budget. And when difficulties do arise, the project manager's focus has to be on keeping ACWP and BCWP closely aligned to minimize variances. Any deviations from the BCWS have to be addressed and corrected immediately so that deviations do not accumulate and that large deviations shown in Figure 10.3 do not occur. The longer deviations keep piling up, the more difficult it will be to correct them. As shown in Figure 10.3, deviations cause schedules to exceed their planned durations, and costs exceed their budgeted funds. The project manager's attention has to be on the project's cost and schedule performance from the very first day on.

AI technologies can assist in reporting cost and schedule performances on a weekly and daily basis, catch the dynamics of variances, and detect adverse trends on a day-to-day or at least a week-to-week basis as feasible.

Mastering Project Management – A key to Success[9]

Introduction

In the dynamic landscape of today's business world, achieving success often hinges on the ability to effectively manage projects. Whether one is launching a new product, implementing a strategic initiative, or overhauling internal processes, mastering project management is a crucial skill that can make or break your endeavors. As is illustrated below, a well-structured and well-managed project can disclose hidden risks and prevent failures.

I'd like to share the fundamental steps that I implement to ensure successful project management. To start, I'd like to share a moment that drives this point home.

I was preparing for a major overhaul of our entire organization's technology infrastructure. It's worth noting that we had over 30 locations, and our focus included, but was not limited to, the following: servers, storage, bandwidth, wireless, switches, data center redundancy, power, cooling, and backups. A team member commented that having one, two, or maybe even three major projects at once was possibly reasonable, but having 12 projects simultaneously was not realistic. However, we successfully completed all 12 projects on time and under budget. Here is how.

Step 1: Define Objectives

The first step in any successful project is a crystal-clear understanding of your objectives. What are you trying to achieve, and how does it align with your overall organizational goals? Are you addressing a safety concern? Are you improving productivity or performance? Are you increasing revenue? Defining your objectives sets the foundation for the entire project, guiding decision-making, resource allocation, and team efforts. This initial step is about creating a roadmap that will serve as a reference point throughout the project lifecycle.

---

[9] Frank Schlueter, Director Education Technology and Information Services, Glendale Unified School District, Glendale, California

Successful project managers understand the importance of collaboration during this phase. Engaging key stakeholders, gathering diverse perspectives, and fostering open communication can lead to a more nuanced and accurate definition of project objectives. By involving all relevant parties from the outset, you increase the likelihood of building a shared vision that resonates with everyone involved.

Step 2: Develop a Thorough Plan

With clearly defined objectives in hand, the next crucial step is to create a detailed project plan. This plan should outline the scope of the project, timelines, resource requirements, and risk mitigation strategies. The project plan serves as a dynamic document, evolving as the project progresses and new information becomes available.

Creating a Work Breakdown Structure (WBS) is an important step. Breaking down the project into manageable tasks and assigning responsibilities to team members helps maintain accountability and ensures that progress aligns with the established timelines. Establishing key milestones and performance indicators allows for ongoing evaluation of the project's success and enables timely adjustments to address any deviations from the plan. Gantt charts can be as simple or complex as you want to serve your needs.

Step 3: Assemble the Right Team

A successful project is not solely about tasks and timelines; it's about the people driving it forward. Assembling the right team is a critical step that involves understanding the unique skills, strengths, and expertise each team member brings to the table. Effective project managers recognize the importance of creating a cohesive and collaborative team culture.

Team members should be aligned with the project's objectives, and their roles and responsibilities should be clearly defined. Foster an environment that encourages open communication, collaboration, and innovation. A well-balanced team can navigate challenges more effectively, leveraging diverse perspectives to find creative solutions.

Step 4: Execute

Execution is where the rubber meets the road. With a well-defined plan and a capable team, it's time to put the project into action. Effective project managers monitor progress closely, address issues promptly, and adapt to unforeseen challenges. Regular communication, both within the team and with stakeholders, is essential to ensure everyone is on the same page and aligned with the project's goals.

Continuous evaluation during the execution phase allows for real-time adjustments, ensuring that the project stays on course. Flexibility and adaptability are key traits, enabling project managers to navigate uncertainties and capitalize on emerging opportunities.

Step 5: Assess and Learn

The final step in the project management journey is evaluation and learning. Once the project is completed, conduct a thorough assessment of its success against the defined objectives. Analyze what worked well and what could be improved. This phase provides valuable insights that can be applied to future projects, contributing to a culture of continuous improvement.

Encourage open feedback from the team and stakeholders, fostering a learning environment that values both successes and failures as opportunities for growth. Documenting lessons learned ensures that the knowledge gained from each project becomes part of the organizational knowledge base, enriching future endeavors.

Mastering project management is an indispensable skill on the path to success. Successful project management is not a one-time achievement but an ongoing journey of refinement and improvement. Embrace each project as an opportunity to hone your skills, empower your team, and lead your organization toward continued success.

New Product Development (NPD)

The industry's quest for critical success factors has resulted in a number of success-critical insights and practices. While MS-EXCEL and

MS-PROJECT are important data-manipulating and process-managing software tools for analyzing and managing innovations, New Product Development (NPD) principles form an umbrella for all product development related activities.

The New Product Development Association (NPDA) [27] cites eight critical success drivers. The factors range from product superiority through crisp product definition and development speed to thorough testing and verification of the product. NPDA's suggestion is well in line with the author's proposed Synergetic Five-Step Innovation Process and adds important aspects to the process. [27]

Open Innovation

The notion that internal R&D is the sole source of innovation does not hold anymore. External events and trends have long been recognized as a main source of new knowledge and ideas. Open innovation is facilitated through several vehicles, including corporate offices in innovation-prone regional centers, licenses, acquisitions, and corporate ventures.

New product and corresponding investment ideas can come from a number of sources, including conferences, investor meetings, market observations, and from an understanding of the capability and opportunity of new technologies. [27] Being aware of technologies derived from nanomaterials and nano techniques in the broadest sense can be inspiring. As digital technologies, such as AI, process mining, digital twin etc., are rapidly emerging and change the landscape, a sound awareness of the new landscape and its opportunities may be the richest source of ideas for new products and services.

Smart Products

The availability of artificial intelligence has accelerated the vision for and the development of smart products. Smart products monitor and adjust their internal processes and have the built-in capacity to make adjustments to maintain optimal operation.[27] Other products, like those related to heavy industrial machinery, may report back to the supplier about its performance and send alerts for maintenance requirements before breakdown occurs.

Entire enterprises are being integrated into one major network of decisive business components. The entire process from supply chain through internal operations to product distribution is being integrated into one interconnected system that manages itself. These cyber-physical systems (CPS) manage daily business and planning tasks automatically.

Managing Innovation Paradoxes

Organizations are facing several paradoxes and have to find ways to manage them. These paradoxes stem from personal traits and relate back to the personnel's basic interests, long-term engagement, and responsibilities, and to the individual's perception of personal goals in life. [27]

The division of labor results in the circumstance that people are in charge of selected business aspects. One group's goal may be to maximize, i.e., exploit, the efficiency and use of resources in order to maximize sales and product production. These "exploiters" want to enhance ongoing business activities. Another group may be in charge of exploring new business opportunities. They want to look for new domains for the enterprise. The exploiters and the explorers typically clash in strategy and related budget discussions. Management has to be aware of the contradictory internal forces and develop the expertise to manage them smoothly.

Another paradox exists that can reside within an individual. Developing ideas for new products and services is not an easy task. An idea is the result of tremendous efforts and deep insight into certain issues and open questions. This circumstance can result in strong pride and ownership feelings and an attitude toward defense rather than open discussions and a willingness to accept criticism. In that case, some underlying adverse factors can be overlooked. The resulting projects can be failure prone. Those projects may be characterized as a "wolf in sheep clothing," where the wolf will shake off his sheep's clothing at a critical circumstance and lead the project into catastrophic failure that may cost lives.

Chapter 11
Traffic Flow Analysis and Design for Mass Mobility

Disruptive Logical-Mathematical Intelligence
Revolutionizing the Transportation Industry

Introduction

The problem-solving power of the Synergetic Five-Step Innovation Process (SFSIP) with a creative Logical-Mathematical Intelligence (LMI) at its center shall be illustrated by addressing worldwide issues posed by traffic congestions on freeways and county roadways. To gain actionable insight into the society's trend toward mass mobility needs and into the complex dynamics of traffic flow and its worsening trend, the innovative SFSIP and LMI methodologies shall be applied to create practical solutions for the benefit of the public, the transportation industry, and the society as a whole.

Consistent with the SFSIP process, research of existing literature addresses society's transportation needs today and in the near future. Research of existing literature shall explore the state-of-the-art in traffic flow analysis and the characterization of dominating traffic flow phenomena. Existing LMI shall be reviewed critically, and, if necessary, new LMI methodologies shall be developed with the objective to identify new solutions for the transportation industry.

Process optimization is an approach where the basic process remains unchanged and where the least adverse procedure is identified and implemented. This traditional approach oftentimes overlooks the opportunity for completely different conflict-free solutions. Rich opportunities are particularly offered by currently emerging digital technologies. In the final step of the SFSIP process, the novel LMI shall be applied to explore the envisioned solutions in all details and confirm their practicality and their societal benefits.

The SFSIP approach reaches deeper and further than traditional process optimizations. The SFSIP approach aids in disclosing revolutionary solutions feasible with emerging digital technologies, such as AI, IoT, digital twin, machine learning and high-speed computing technologies.

Challenges of the Global Transportation Industry

A critical qualitative review of ongoing traffic dynamics reveals that the traffic's own inherent phenomena and significant external factors create a confusing labyrinth of phenomena, which cloud solution-finding processes. To clear the path to ideal conflict-free solutions, three issues shall be addressed.

Issue One - Prevailing Traffic Flow Management

The approach to exploring solutions for failing traffic flow on freeways and county roads is typically focused on predicting traffic congestions and guiding vehicles to circumvent the congested areas by changing their routes. For that purpose, traffic planners propose to utilize live data and apply network analysis with the objective of designing and implementing adaptive traffic management systems for entire road networks. Some planners propose novel cooperative vehicular systems based on vehicle-to-vehicle (V2V) communication to detect traffic congestions using fuzzy logic. The techniques applied are designed to detect incipient traffic congestions and minimize their impact by spreading the traffic throughout the network.

It is believed that ongoing traffic management systems designed to circumvent detected congestion areas and, thus, minimize congestion, can only be intermediate solutions that cannot have much value in the long run. Smart systems in the digital age must represent completely new approaches that take the demanding phenomena of future mass mobility into full account and take full advantage of pioneering technology-based opportunities. Lane capacities of freeways and county roads must match and exceed mass mobility requirements and prevent the formation of traffic congestions in the first place.

Issue Two - Growth Trends of Population Centers

Populated regions are defined in terms of population density. Cities make highly dense regions, where people live close together, as is illustrated by the charts in Figures 11.1 and 11.2.[51] Although people have access to various forms of public transportation, city streets are crowded with automobiles, and traffic is controlled by traffic lights. Cities are surrounded by large, less densely populated county regions, where most people live in single-family homes. Rural areas feature plenty of open space utilized for agricultural purposes.

Figure 11.1 Population, GDP and Energy Consumption [51]			
	2023	2050	Ratio (%) 2050/23
World Population, B	7.8	12.6	161%
Urban Population, B	4.4	8.8	200%
Urban Population, %	56%	70%	124%
GDP, % of World	80%		
Energy, % of World	65%		
Emissions, % of World	70%		
Land Use, % of 2023	100%		300%

City and county areas combine to form major population centers. Because of the less dense population in county areas, public transportation is less developed, and people depend mostly on their own automobiles to commute to and from work and for their personal needs and leisure travels. In spite of multi-lane county roads and freeways, traffic has outgrown road expansion options in many regions around the world.

Today, 4.4 billion people live in county environments, which amounts to more than 50% of the world's current 7.8-billion population. According to the World Bank, a strong urban growth trend is expected to continue, with its population more than doubling its current size by 2050, at which point nearly 7 out of 10 people will live in cities and their surrounding county areas. (Figures 11.1 and 11.2)

The World Bank [51] reports that more than 80% of global GDP is generated in cities and their immediate county areas. Consequently, the World Bank concludes that urbanization will contribute substantially to societies' sustainable growth. On the other hand, cities and counties represent two-thirds of global energy consumption and account for more than 70% of greenhouse gas emissions. Consequently, the continued growth along with concurrent increases in productivity and innovation must be managed responsibly. [51]

The progression and scale of population growth in cities and counties brings challenges, such as meeting accelerated demand for affordable housing and employment, viable infrastructure including transport systems, and other basic utility services. The nearly 1 billion urban poor, who live in informal settlements to be near income providing opportunities, must be included in the planning process. These challenges do not go away by themselves; in fact, they will get worse with time unless effective formal solutions are identified and implemented.

Planning is an important chore because, once a community's infrastructure has been built, its physical form and land-use patterns are locked-in for generations, leading to unsustainable sprawl. Such sprawl puts pressure on land and natural resources, resulting in undesirable outcomes.

The expansion of land consumption outpaces population growth by as much as 50%, which is expected to add 1.2 million km² of new urban built-up area to the world by 2030. [51] If the World Bank's 2030 projections for urban land consumption were extrapolated to the year 2050, where the rise in urban population is forecasted at 200%, urban land consumption may grow by 300% compared to 2023. [60]

Building cities that "work," i.e., green, resilient and inclusive, requires intensive policy coordination and investment choices. National and local governments have an important role to play to act now, to shape the future of their development, and to create opportunities for all. [51,60]

Ongoing planning efforts include programs designed to expand infrastructures consistent with the various needs posed by housing, education and employment, health, transport, utilities, sanitation and waste,

	Figure 11.2 GDP and Revenue in Urban Areas		
	US	Europe	World
Urban Population, % of World	80.0%	60.0%	56.0%
GDP of urban Areas	84.0%	65.0%	80.0%
5 Top Tech Corp's Revenue			70.0%
Apple's Revenue			85.0%

integration and social cohesion, safety and security. [60]

Advanced technologies, such as AI, IoT, process mining, digital twin and emerging nanotechnologies, play major roles in the planning effort. Exceptional opportunities and the pertaining uncertainties affect the planners' vision and problem resolution capacity. Members of international planning communities should mutually agree to build a shared vision so that the cities of the future can indeed become the kind of cities that we really need and want.

Stanford University's Program on Disruptive Technology and Digital Cities (DTDC) is a prime example of academia's and industry's awareness of challenges facing urban and suburban developments in the digital age. [48] Stanford's DTGC Program considers disruptive technologies and new business models spanning healthcare, financial services, transportation, construction, sustainability, energy, advanced materials, data analytics, media, and entertainment. Stanford's DFDC program provides its members with active guidance on technology identification and how to transform disruptive technology into new opportunities for growth. The program's approach offers a new and interdisciplinary perspective on technology by combining business modeling expertise and technology research. [48] Stanford University's statistics expand the World Forum's and the World Bank's observations and address the urban development challenges of the future.

The statistics imply an exceptional worldwide trend toward exponential growth in urban and suburban population centers in the very near future. In light of land limitations, a shift from horizontal to vertical suburban planning is gaining momentum. The corresponding rise in population density is expected to raise traffic density as well and will worsen ongoing traffic flow dilemma.

3-D planning must not be limited to buildings for various community needs; it must also address transportation. As we approach mass mobility and want to fulfill the public's affinity toward personal vehicles, technology-based transportation options must be addressed without delay.

Because many county infrastructures are locked-in so that physical expansion of existing infrastructures is not possible anymore, solutions must be found that raise the transportation industry's productivity without expanding physical infrastructures.

Issue Three - Freeway Traffic Flow Dynamics, Puzzling Phenomena

Traffic congestions and slow traffic flow used to occur when lanes were blocked by physical obstructions, such as traffic accidents and road maintenance work. When drivers had reached the front of the congestion and were able to pull ahead, they could catch a glimpse of what had caused the traffic delay. Advanced automotive technologies, improved infrastructures, and more disciplined driving significantly reduced physical road obstructions. But now a different kind of traffic flow obstruction in the form of slow-moving traffic jams has emerged. Drivers pulling out in front of slow-moving traffic jams look for an obstruction that may have caused the traffic delay, but there is none. The new form of traffic jams appears as a puzzling phenomenon. It seems that slow-moving traffic jams are caused by "virtual road obstructions" that are initiated by inherent traffic flow phenomena, which may be instigated by human driving behavior.

Anticipating that the growth in population centers will continue to raise the frequency and the size of slow-moving traffic jams, it was concluded that traffic's inherent affinity toward slow-moving traffic jams needs to be fully understood in order to devise a traffic flow management solution for today and for the future.

During an extensive qualitative research of traffic phenomena, approximately 100 international publications issued during the past 60 years were reviewed. [53,54] The search for information regarding the puzzling slow-moving traffic jam phenomenon led to two decisive experiments conducted by research teams at Massachusetts Institute of Technology and Bristol University in the UK.

In 2008, an MIT team assembled an electronic traffic flow model to research the slow-moving traffic jam phenomenon.[31] The team designed a traffic flow simulation supported by Bernoulli's laws of fluid flow mechanics.[62] The team assigned a fluid-type viscosity to the stream of vehicles and applied computational fluid dynamics (CFD) to explore the formation of slow-moving traffic jams caused by traffic overloads, not by physical road obstructions.

MIT's CFD model illustrates the slow-moving traffic jam phenomenon caused by traffic overloads. At saturated traffic, i.e., when every vehicle has the space that its driver needs to maintain the assigned velocity, the traffic flows smoothly without forming slow-moving traffic jams. Under overload conditions, MIT's CFD model illustrates that a certain number of vehicles are indeed squeezed into a slow-moving traffic jam.

MIT's CFD-based traffic flow simulation model deserves a closer look in terms of its underlying nature. Most importantly, it does display very clearly the formation of slow- moving traffic jams without the presence of any physical road obstruction. Bernoulli's classic fluid flow laws demand that the mass flow is constant even in the presence of a flow obstruction. A traffic overload represents an obstruction in the context of fluid flow behavior. In that case, the CFD laws facilitate that the vehicles representing the overload are pushed into a slow-moving traffic jam. In addition, consistent with Bernoulli's fluid flow laws, the non-effected free-moving vehicles are accelerated to a higher velocity to the extent that the overall mass flow across the entire model track is maintained at the original value. However, since the additional vehicles that form a close-spaced, slow-moving traffic jam, occupy a certain section of the track, the space available for the remaining free-moving vehicles is less than before so that the remaining free-moving vehicles are spaced closer than before. Under human driving conditions, they would proceed at a lower velocity than before, whereas the CFD model law forces them to accelerate. Further, all CFD controlled vehicles behave the same way in the model, where, in reality, human drivers behave very differently. These observations do not distract from the value of MIT's CFD model. In fact, these observations contribute effectively to the search for an ideal conflict-free traffic management solution.

In 2015, a Bristol University team in the UK performed a physical field experiment, where the Bristol team applied MIT's simulated dimensions to a real physical field test. [5] The field test uses a real track, real vehicles, and real drivers. The Bristol field experiment confirmed MIT's formation of slow-moving traffic jams caused by traffic overload

157

and not by physical obstructions. There is one significant difference, though. While MIT's model displays a single slow-moving traffic jam that is very stable and does not change in size as long as the traffic overload definition is not changed, the Bristol field test with real drivers displays continuous change in traffic jam size. Since the group of real drivers resembles a wide spectrum of driver behavior ranging from very cautious to very aggressive, the jam size varies constantly depending on which drivers are temporarily in the slow-moving traffic jam and which are not. Further, the jam formation does not occur as a single jam; in fact, it occurs randomly at more than one position within the group of vehicles on the track. The sum of vehicles trapped in slow-moving traffic jams tends to be much larger than the traffic overload. This observation indicates that, under realistic human driving conditions, the slow-moving traffic jams include a number of vehicles that is larger than the number of overload vehicles. Thus, when traffic overload beyond traffic saturation causes the formation of slow-moving traffic jams, there is a multiplier effect at work.

The MIT and Bristol models provide decisive information for moving forward to specify an ideal conflict-free solution.

Freeway Traffic Flow Analysis and Design

The MIT and Bristol models explain the formation of puzzling slow-moving traffic jams. The culprit is traffic overload, where the overload condition is caused by human drivers' natural need to maintain comfortable following distances.

The goal is to eliminate the formation of slow-moving traffic jams, which may require that the human factor is eliminated. Such a solution appears to be feasible because of emerging digital technologies. Consequently, the ideal conflict-free solution is conceived as a fully automated traffic management system that eliminates the human factor. Since the human factor will remain involved during the transition to the ultimate solution, the ideal solution must include a manageable transition, where human driving habits are smoothly transitioned to the final goal.

This concludes steps one and two of the Synergetic Five-Step Innovation Process, and steps three and four, i.e., quantitative analysis using Logical-Mathematical Intelligence, can commence.

Creating Insightful Actionable Intelligence

The process solution accomplished in the high data-traffic telecommunications space is a prime example for industries that face a fast-paced growth in terms of traffic loads. Society's emerging need for mass wireless telecommunications was recognized in the 1980s by visionary entrepreneurs familiar with the limited availability of radio frequencies. The capacity of the then available physical infrastructure was too limited for the anticipated mass communication age. Entrepreneurs applied LMI in electronics, defined an ideal conflict-free solution, and solved the problem by creating a technology-based virtual infrastructure that was superimposed on the limited physical infrastructure and raised the capacity significantly. The virtual infrastructures are enabled by nano-time increments of human speech and have become known as Code Division and Time Division Multiple Access (CDMA, TDMA).

The development of freeway and county traffic management designs for mass mobility in the digital age has to be accomplished comparable to the transformation accomplished in mass telecommunications. The envisioned ideal conflict-free solution, i.e., a fully automated traffic management system, may require a virtual infrastructure that can be superimposed on the existing physical infrastructure provided by existing freeways and county roads.

There is a major difference, though, between mass telecommunications and mass mobility. In mass telecommunications, the users' voices and corresponding data are the objects that are carried at high speed to their destinations. The users are simply acting as system-external users of the high-speed and high-capacity CDMA and TDMA systems. In contrast, in mass mobility systems, the users are system-internal participants. The users are the objects that need to be moved at high speed and in mass configurations to their desired destinations. This circumstance indicates that a mass mobility solution must not only provide an ideal conflict-free end state, but it must also provide for a gradual introduction to gradually build the consumer's confidence in the ultimate solution.

Human Driving Behavior

According to the solution finding sequence defined by the SFSIP process, a suitable Logical-Mathematical Intelligence must be defined to assess the functionality and feasibility of the envisioned ideal conflict-free solution for the transportation industry.

The traffic flow conditions evidenced by the MIT and Bristol studies are of a dynamic nature. In an effort to explore those dynamics further, it was found that the large volume of existing traffic data [53, 54] represents single-parameter data, which imply static traffic flow conditions and, thus, do not lend themselves to dynamic traffic flow analysis. Based on MIT's and Bristol's findings, it was felt that dynamic traffic flow data were needed that could disclose further relationships between traffic flow velocity, traffic density, driver behavior and the traffic jam formation dynamics.

It was observed that traditional methodologies applied from the sideline of traffic flow could not reach beyond single isolated data. The needed relationships could have been attained by engaging IoT, global positioning systems (GPS), and advanced navigation systems (ANS). To accelerate the research project and to minimize its cost, an innovative data acquisition approach was devised. The data acquisition had to be moved from the sideline to the inside of the traffic scene.

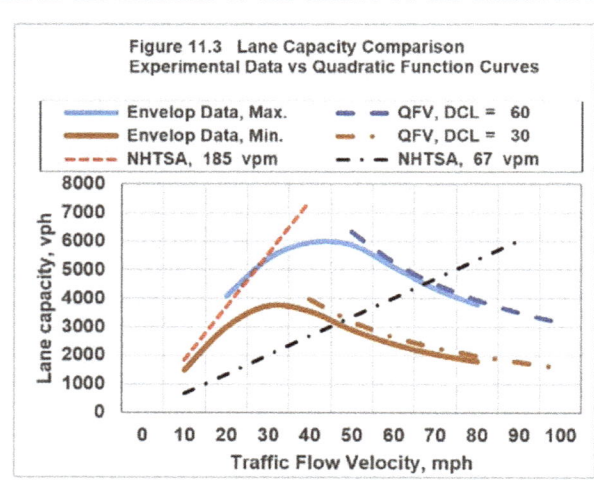

Figure 11.3 Lane Capacity Comparison Experimental Data vs Quadratic Function Curves

The research team traveled along with the freeway traffic and recorded their observations of practiced following distances and corresponding velocities. [45] The data thus obtained was organized into maximum and minimum center-to-center intervehicle spacing for selected velocities as shown in Figure 11.3.

It was found that the maximum and minimum data resembled enveloping curves embracing all driving habits. The enveloping curves resemble an exponential characteristic, where lane capacity tends to be reversely proportional to the traffic flow velocity, as is illustrated by the solid curves of the chart in Figures 11.3.

The chart also includes the maxim traffic density of 185 vehicles per mile (vpm), at which traffic breakdown occurs. [53]

The National Highway Traffic Safety Administration (NHTSA) reports that traffic flow instability tends to affect traffic flow at vehicle densities larger than 67 vpm. The experimental data in Figure 11.3 indicates, though, that the 67-vpm mark applies to cautious drivers at velocities of approximately 45 mph. For aggressive drivers, the traffic breakdown may occur at approximately 100 vpm and at velocities of 55 mph.

Driver Comfort Level (DCL) - A Logical-Mathematical Intelligence

Physics laws for the energy content of a moving body specify that the energy of an object in motion is proportional to the velocity squared. This implies that the breaking distance of vehicles is proportional to the velocity squared. Drivers may not be familiar with this natural law, but they experience disproportionately longer breaking distances at higher velocities. As a result, human drivers tend to expand their following distance at higher velocities. Thus, traffic density tends to be lower at higher velocities and vice versa. Introducing a factor for the human drivers' comfort level (Driver Comfort Level, DCL[10], patents [42,43,44]), the intervehicle spacing formula can be derived from the energy law.:

$$\text{Intervehicle Spacing} \quad = \quad 1/\text{DCL} \ * \ \text{Velocity}^2 \, [\text{ft}]$$

The DCL is a novel engineering-type characterization of human driving behavior in terms of two major traffic parameters, i.e., velocity and intervehicle spacing. Using the experimental dual-parameter data [45], the DCL parameter can be determined for the data enveloping curves:

$$\text{DCL} = 1 / \text{Intervehicle Spacing} * \text{Velocity}^2 \quad [\text{mph}^2 / \text{ft}]$$

Knowing a group of drivers' average DCL value, saturated traffic density can be calculated for a one-mile section.

$$\text{Traffic Density} = 5,280 * \text{DCL} / \text{Velocity}^2 \quad [\text{vpm}]$$

According to Bruce Greenshields' basic mathematical law, lane capacity is the product of traffic density and velocity [66,67]

$$\text{Lane Capacity} = \text{Traffic Density} * \text{Velocity} \; [\text{vph}]$$

Inserting the equation for traffic density, delivers the Quadratic Function of Velocity (QFV) correlation for Lane Capacity as it evolves for human driving conditions.

$$\text{Lane Capacity} = 5,280 * \text{DCL} / \text{Velocity} \quad [\text{vph}]$$

Using maximum and minimum DCLs of 60 and 30 mph^2/ft, respectively, lane capacity curves per the above QFV were entered into the Chart of Figure 11.3. The correlation achieves near-perfect matches with the experimentally obtained enveloping curves.

At each velocity, drivers apply their DCL-dependent intervehicle spacing so that the maximum and minimum data-enveloping curves in Figure 11.3 represent traffic saturation. As additional vehicles enter the lane, the human-driven vehicles reduce their velocities in response to the reduced intervehicle spacing. This response provides the needed increase in lane capacity and maintains the traffic saturation. Once the density limit, i.e., approximately 67 to 100 vpm per the NHTSA [53], is approached, the traffic become unstable and breaks down completely at approximately 185 vpm.

Human driving habits cause a serious conflict. Additional traffic loads are accommodated by reducing the traffic flow velocity. Additional vehicles are accommodated at the cost of every vehicle's speed and causes frustrating travel delays. As the traffic flow velocity continues to be reduced there is a natural limit at which the traffic breaks down. This represents a serious system instability that needs to be cured.

Dynamics of Slow-Moving Traffic Jams

The MIT and Bristol University traffic flow studies involving computational fluid dynamics (CFD) and physical road tests, respectively, reveal that the formation of close-spaced vehicle clusters, i.e., slow-moving traffic jams, may be governed by natural laws. The fluid dynamics laws utilized by MIT's research team illustrate the traffic jam formation in a very disciplined and transparent manner. But the CFD methodology's principle of constant mass flow does not yield a realistic combination of spacing and velocity for the free-moving vehicles. It appears that a practical mass and space balance for the integral system of jammed and free-moving vehicles should be explored. This observation inspired this author to apply mass and space balance principles and figure out if the results could match the results achieved by the Bristol University's field test with real vehicles and real drivers.

The mass balance on a one-mile lane section with free-moving vehicles (FV) and slow-moving jammed vehicles (JV) can be defined as the sum of both types of vehicles.

Total Vehicles (TV)   =   Free Vehicles (FV) + Jammed Vehicles (JV)

For the space balance, the intervehicle spacing of free-moving vehicles (FVS) and jammed vehicles (JVS) must be defined. The drivers of the free-moving vehicles will want to maintain an intervehicle spacing depending on their DCL and the desired velocity consistent with the pertaining speed limited.

FV Spacing (FVS)   = $\text{Velocity}^2 / \text{DCL}$ [ft]

The jammed vehicles are traveling at a varying low velocity between zero and 10 or 15 mph, which has no effect on the space balance assessment. The intervehicle spacing varies in an undefinable velocity-independent manner ranging from approximately 20 to 25 ft. An empirical value like, for example 23 ft, can be used as average spacing.

$$\text{JV Spacing (JVS)} \quad = \quad 23 \quad [ft]$$

The intervehicle spacing definitions are used to formulate the space balance condition for a one-mile lane section.

$$FV * FVS \ + \ JV * JVS \quad = \quad 5{,}280 \quad [ft]$$

The mass and space balance conditions can be solved for the number of jammed vehicles.

$$JV \quad = \quad (TV * V^2/DCL - 5280) / (V^2/DCL - JVS)$$

or

$$JV \quad = \quad (TV * FVS - 5280) / (FVS - JVS)$$

The number of jammed vehicles (JV) is determined as a function of the total number of vehicles per one-mile section, their preferred intervehicle spacing (FVS) and the jammed vehicles' average intervehicle spacing (JVS).

The JV formula represents true logical-mathematical intelligence. When the total number of vehicles (FV) multiplied by their preferred spacing (FVS) equals the length of the one-mile road section (5,280 ft), the traffic has just reached its saturation and, theoretically, the number of jammed vehicles is zero. When the number of vehicles per one-mile section exceeds the saturation condition, the JV formula determines the excess space (TV * FWS – 5,280) and divides it by the difference in intervehicle spacing between the two types of vehicles (FVS-JVS). The later expression represents the space savings accomplished when one FV is pushed into the traffic jam. Consequently, the excess space claimed by all vehicles under free-moving conditions is converted into the number of jammed vehicles to accomplish a balanced traffic condition. With this logical justification for the validity of the JV correlation, the correlation can be considered a logical-mathematical intelligence of traffic flow theory.

The JV formula can be utilized to determine the traffic's sensitivity to overload conditions and its affinity to form slow-moving traffic jams of various sizes. The blue curve in Figure 11.4 represents a scenario in which a saturated traffic condition is reached with 100 vehicles per mile traveling at 56 mph and maintaining intervehicle spacing of 52.8 ft. When 20 additional vehicles enter that lane, which amounts to an increase of 20%, approximately 30% of all 120 vehicles will be pushed into slow-moving traffic jams. In other words, 20 additional vehicles cause a traffic jam size of 36 vehicles. The traffic jam formation involves a multiplier effect, which amounts to 1.8 for the blue curve.

The red curve resembles a traffic condition, where the saturation is reached with 160 vehicles traveling at 44 mph and maintaining intervehicle spacing of 33.0 ft. Adding 20 vehicles would raise the total number of vehicles to 180, a 13% increase, and would cause 40% of all vehicles, i.e., 72 vehicles, to be squeezed into a slow-moving traffic jam. The traffic jam multiplier in that scenario amounts to 3.6.

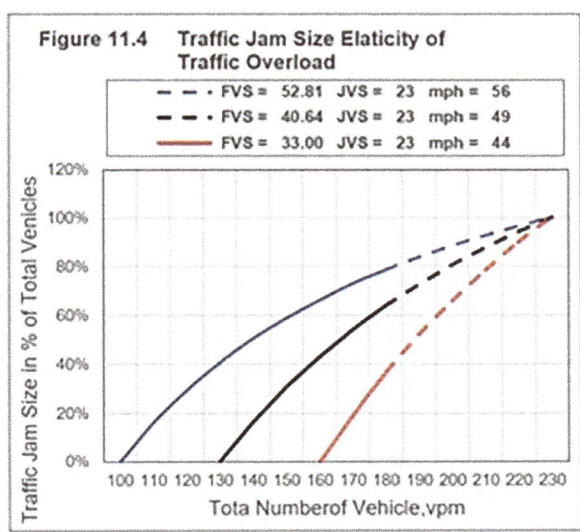

Figure 11.4   Traffic Jam Size Elaticity of Traffic Overload

The numerical results for traffic jam sizes, where the number of jammed vehicles is larger than the traffic overload, is consistent with the conclusions gained from observing Bristol University's physical experiments.

The comparison of low-density traffic conditions (the blue curve of Figure 11.4) with high-density conditions (the red curve of Figure 11.4) reveals that high-density saturated traffic conditions feature a stronger jam elasticity to traffic overloads. The traffic jam size multiplier is shown in Figure 11.5 as a function of traffic density in terms of vehicles per mile (vpm). The chart illustrates that the traffic jam size multiplier rises steeply as the traffic density rises.

When this observation is applied to the traffic flow dynamics, we can conclude that saturated high-velocity traffic conditions, which typically are associated with low traffic densities, feature low risks for slow-moving traffic jams, i.e., additional vehicles entering the lane will most likely be accommodated by velocity reduction and an associated increase in lane capacity. In contrast, saturated low-velocity traffic conditions, which typically are associated with high traffic densities, feature high risks for traffic jam formation. The real-life traffic dynamics represent a combination of the two phenomena discussed here.

The Logical-Mathematical Intelligence practiced above leads to the observation that slow moving traffic jams can indeed occur at any intermediate velocities and densities. The likelihood of traffic jams increases as the traffic density rises. Thus, the peak lane capacities indicated by the curves in Figure 11.3 are not likely to occur, and if they do, they represent highly unstable traffic flow conditions.

In summary, there are two phenomena occurring due to human driving behavior. Initially, additional vehicles entering an unsaturated traffic condition will reduce the intervehicle spacing and raise the traffic density accordingly without affecting the traffic flow velocity. Once traffic saturation is attained, further vehicles entering the lane will either cause a reduction in velocity or will cause the

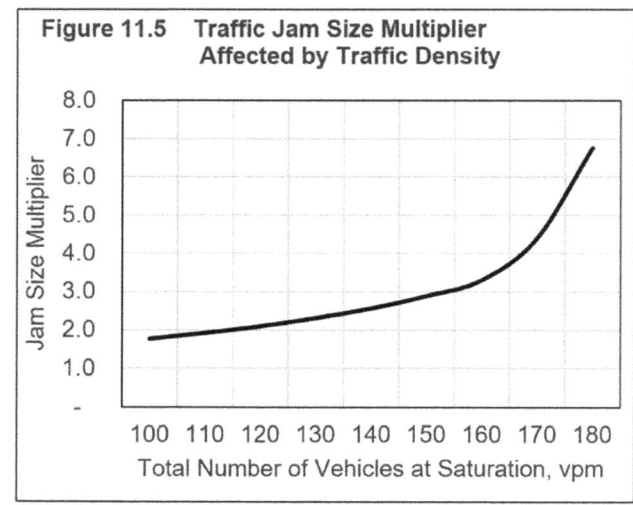

Figure 11.5   Traffic Jam Size Multiplier Affected by Traffic Density

initiation of slow-moving traffic jams. For the researcher, who is striving for an ideal conflict-free solution, it is not important to know which of the

166

two phenomena occurs and when; it is only important to understand both phenomena and to define an ideal conflict-free solution, that will resolve both traffic flow phenomena.

Definition of Ideal and Conflict-Free Traffic Flow

The Logical-Mathematical Intelligence practiced above leads to the conclusion that traffic improvement efforts to accommodate our society's thirst for mobility must aim at the elimination of traffic jams.

Figure 11.5 illustrates that traffic flow is particularly vulnerable to breakdown at elevated traffic density conditions. In contrast, Greenshields' fundamental correlation of lane density specifies that high lane capacities require high traffic density and unrestricted traffic flow velocity. Thus, human controlled traffic flow is in dire conflict with high lane capacity as demanded by society's current and future mobility.

With that insight, the traffic's ideal conflict-free solution is obvious. It confirms that slow-moving traffic jams must be prevented, and unrestricted traffic densities and velocities must be enabled.

Freeway Lane Capacity to Meet Mobility Requirements

Design of conflict-free traffic flow on freeways requires fundamental deviations from traditional thinking. As was discussed above, high lane capacities require high traffic density that is not limited by velocity. It shall only be limited by physical intervehicle spacing limitations. With that ideal goal in mind, lane capacity must be directly proportional to the product of traffic density and traffic velocity according to the fundamental lane capacity functionality defined by Bruce Greenshields in 1953. [66,67]

$$\text{Lane Capacity} = \quad \text{Density} * \text{Velocity} \quad [\text{vph}]$$

Above formula is illustrated in Figure 11.6. The green line represents the proposed automated freeway traffic management system using a constant traffic density of, for example, 200 vehicles per mile (vpm). At 100 mph and a lane capacity of 20,000 vph, one can consider a high-speed tunnel condition where vehicles can accelerate beyond 100 mph under constant interval and correspondingly lower density conditions, so that the lane capacity remains constant. At the end of the high-speed tunnel, the velocity is reduced to the 100-mph level, and the density of 200 vpm is re-established accordingly.

The question arises of how the transition from today's human controlled traffic flow to the automated traffic flow may be structured. The key to this question lies in the novel driver comfort level (DCL). As shown in Figure 11.6, an aggressive driver's comfort of 80 mph²/ft can yield a lane capacity of approximately 5,000 vph at a velocity of 85 mph. Assisting drivers to achieve DCLs in the 150 mph²/ft range can yield lane capacities in the 10,000 vph range. This improvement in DCL performance must continue so that at one point the transition to a fully automated traffic management system can be executed. In addition, a low vehicle density can be implemented initially to ease the drivers' transformation into the transportation mode of the digital age.

There are two key performance parameters, i.e., the driver comfort level and the vehicle density, that can be considered to gradually accomplish the drivers' transformation. As drivers become comfortable with digital technologies and automation as a whole, the vehicle density, i.e., DCL, and the traffic flow velocity can both be gradually raised

The fully automated Freeway Traffic Management is a high-performing Multi-Lane Traffic Management

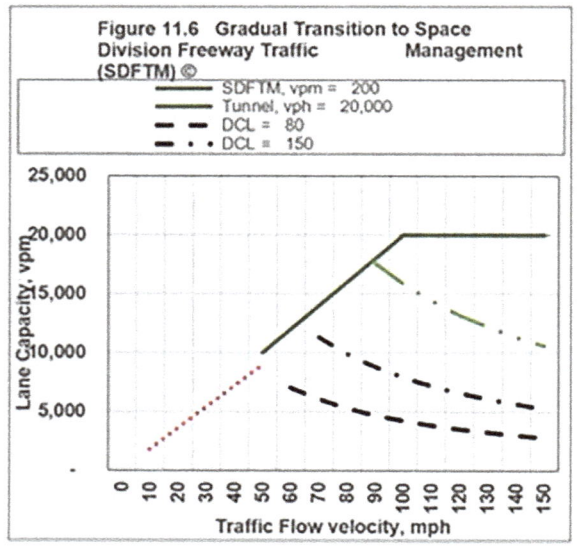

Figure 11.6 Gradual Transition to Space Division Freeway Traffic Management (SDFTM) ©

— SDFTM, vpm = 200
— Tunnel, vph = 20,000
– – DCL = 80
– · · DCL = 150

System for Platoons of Autonomous Vehicles.[11] The design details are described in a patent application [42] published on November 2, 2023 by the US-PTO.

Freeways: Space Division Freeway Traffic Management (SDFTM)[12]

Stability Requirements for High-Density and High-Velocity Traffic Flow

The challenges posed by automated traffic management systems, which allow high-density and high-velocity traffic flow, are best illustrated by taking a brief look at initial field experiments configured to explore and demonstrate the performance of automated systems.

The complexity of traffic flow dynamics leads traffic system designers to emphasize adaptivity as a key component of their automated traffic flow design concepts. They consider it important to allow vehicles to be responsive to changes in traffic loads. Further, it is believed that the commercial transport industry involving large trucks piloted by professional drivers may be the first logical choice for practicing automated traffic management systems.

Several successful tests have been performed with a group of large 16-wheeler trucks, equipped with automated traffic management systems, including automated management of their following distance. During several field tests, human-driven sedans were instructed to enter the truck convoy between two trucks. The developers observed with satisfaction that their systems responded in a safe manner according to plan. The affected trucks slowed down to reestablish their programmed following distance. Since this automated behavior resembles exactly the initiation of slow-moving traffic jams, the intended adaptive performance may cause a problem for large-scale implementation of adaptive traffic management strategies.

Multi-Lane Traffic Management System for Platoons of Autonomous Vehicles[13]

The thought experiment for high-density and high-velocity traffic flow must include specifications for creating absolute traffic flow stability and must aim to prevent slow-moving traffic jams. Both requirements are not independent of each other; they are integrally combined and, thus, go hand-in-hand.

If one considers the automated high-density and high-velocity traffic flow management systems for a single lane, the system does not pose complex issues. In contrast, if one wanted to practice high-density and high-velocity traffic flow configurations for a dynamic multi-lane freeway system, it becomes a complex system with multiple challenges.

The practical implementation of consistently stable high-density and high-velocity traffic flow conditions requires a feature that allows vehicles to safely change lanes in such a manner that traffic flow stability is maintained. A virtual infrastructure is envisioned in the form of a Space Division Freeway Traffic Management (SDFTM)[14] platform that can enable the intended high-density and high-velocity traffic flow and maintain absolute traffic flow stability. The design of the technology-enabled virtual infrastructure is defined in a patent application. [42,43]

The virtual-infrastructure-based Space Division Freeway Traffic Management (SDFTM) system fulfills the thought experiment for future traffic flow configuration to meet ongoing and future mobility requirements with a broad margin against slow-moving traffic jams and traffic breakdowns. The virtual infrastructure overlays the physical infrastructure to maximize its productivity. At 75 mph a lane capacity of 15,000 vph can become a reality. And as future automotive technologies, including autonomous vehicles, can enable traffic flow velocities of 100 mph and more, the high-speed lane capacities can reach and exceed 20,000 vph. Compared to ongoing traffic jam conditions, lane throughput

170

capacity is increased ten-fold or more without having to expand existing physical infrastructures.

The virtual infrastructure based SDFTM methodology for Platoons of Autonomous Vehicles is not aiming at detecting traffic congestion and leading vehicles to alternate routes to minimize traffic congestions. In contrast, the virtual infrastructure-based SDFTM raises lane capacities multi-fold and prevents congestion prone traffic conditions in the first place.

County Roads: Space Division County Traffic Management (SDCTM)[15]

A Generic Issue Hinders Traffic Flow on Freeways and County Roads Alike

On freeways, we observe that saturated traffic conditions display a large traffic jam elasticity to traffic density, i.e., the number of vehicles pushed into a traffic jam is a multiple of the traffic overload (see Figure 11.4).

The described dynamics of the traffic jam formation are very similar on county roads. On county road systems, the traffic jam syndrome is caused by traffic signals. In both traffic systems, the problem seems to be the same, i.e., the provided lane capacities are insufficient for the society's demand for mobility. On freeways, a natural instability causes saturated traffic to collapse and form slow-moving traffic jams. The phenomenon can be described as a slow-moving virtual red light. On county roads, physical red lights interrupt the traffic flow and drastically reduce the roadway's throughput capacity.

In today's digital age with an abundance of powerful technologies available, the solution for county roads does not lie with a physical expansion of the physical infrastructure. In contrast, road capacities must be raised multifold to meet mass mobility requirements. The solution is similar to that proposed for freeways. A virtual infrastructure must be

---

superimposed on the physical infrastructure of county roads to minimize the red-light interference and to maximize the roadway's productivity.

Green Wave of Traffic on Bidirectional County Roads[16]

The evolution of traffic phenomena on County roads is unique. The original purpose of traffic lights was to control traffic flow with the objective to enable safe traffic flow in all directions across intersections. This arrangement works well when traffic loads are relatively low, and signaled intersections were far apart from each other. As traffic volumes increased and the number of signaled intersections increased accordingly, traffic flow began to suffer. In light of today's strong thirst for mobility, traffic flow on county roads experiences traffic breakdowns in the worst possible ways. Traffic flow becomes a stop-and-go affair, where the traffic moves from red light to red light. This situation causes not only extreme inconveniences and lost time for the drivers and for commercial organizations, but it also causes unacceptable environmental impacts.

The culprit of unacceptable county traffic flow conditions is inappropriate signal control. [44,45] Traffic lights tend to be independently controlled by cross traffic. At most locations, traffic signals are randomly switched according to random cross traffic demands. In today's digital age, where data acquisition, big data analysis, and predictive data analytics are in fashion, adaptive traffic management systems do not yield the capability to eliminate the apparent random control of traffic signals.

Current traffic management systems address the entire network of roads and constantly analyze and manage traffic flow as a whole, which requires an enormous data acquisition system, quick big-data analysis, and constant modifications of traffic signal management. This is a demanding task although it can be effectively handled by today's digital technologies. The question remains, though, how effective the resulting solutions really are. It is observed that current traffic management systems lack transparency and simplicity. They are just as complex as current non-managed systems are. The solutions proposed for county roads do not

---

172

provide effective responses to rising traffic loads and do not respond to the unique requirements of bi-directional roadways. Current traffic management systems achieve only partial improvements.

Being guided by the conflict-free "green wave" concept, which can provide maximum traffic flow, but which, of course, does not work on bi-directional county roads, a virtual infrastructure was considered. The resulting Space Division County Traffic Management (SDCTM)[17] platform was found to achieve a green-wave-type conflict-free traffic flow solution for bi-directional county roadways.

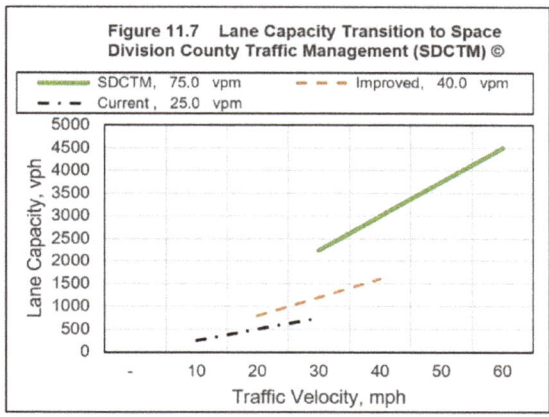

Figure 11.7   Lane Capacity Transition to Space Division County Traffic Management (SDCTM) ©

As shown in Figure 11.7, on county roads, non-managed traffic flow suffers from randomly switched traffic lights and achieves average velocities of 30 mph or less. The resulting lane capacity tends to be less than 1,000 vph. As shown in Figure 11.8, the travel time for a 10-mile trip can amount to 60 minutes or more.

As shown in Figures 11.7 and 11.8, currently available traffic management systems achieve partial improvements, where the majority of traffic signals still interrupt the traffic flow so that the average velocity may still be limited to 40 mph or less. The resulting lane capacity may approach 1,500 vph, and the travel time is improved by approximately 30%. [44]

The proposed SDCTM platform removes all remaining obstacles (solid green lines in Figures 11.7 and 11.8). The proposed virtual-infrastructure-enabled Green Wave of Traffic can generate bi-directional green waves of traffic. Traffic on both sides of a bi-directional county road can travel without interruptions. Lane capacity is raised to the 4,000-vph range or more depending on the specified traffic velocity. Thus, the SDCTM platform raises traffic capacities by a factor of three, and velocities by a factor of 2.5 for a combined productivity improvement of approximately 750%.

The beauty of the proposed Bidirectional Green Wave of Traffic is that the actual velocity is not adversely affected by randomly switching traffic lights so that the traffic can move according to the county's set speed limits. The 10-mile travel time can amount to 10 minutes or less.

The benefits of the solution are multifold. The County Traffic Management System achieves the highest possible lane capacity and the lowest possible travel time. Fuel consumption, air, and noise pollution for neighboring communities will be reduced as well. The system design is transparent, and its performance can be raised further as automotive technologies experience future advances.

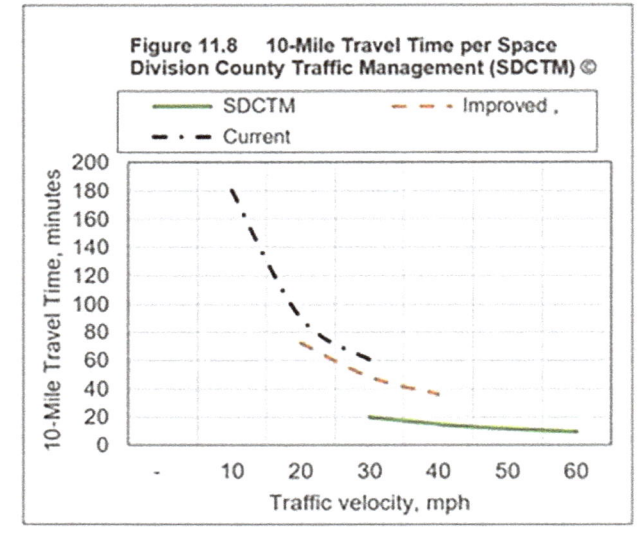

Figure 11.8   10-Mile Travel Time per Space Division County Traffic Management (SDCTM) ©

Implementation and operations costs are minimal compared to other currently promoted solutions. The SDCTM design details and its functional performances are the subject of pending patents filed with the US Patent and Trademark Office. (US-PTO) [42,43,44]

The virtual infrastructure proposed for county roads offers a significant system-inherent advantage. As travel time is significantly reduced, each vehicle claims its space on the road for less time. This will greatly reduce the risk of congestion and will contribute to allowing free traffic flow.

The Space Division County Traffic Management methodology raises the capacity of county roadways well above current rush hour capacity needs. The system offers a traffic management option that exceeds current mobility needs and has the potential for further growth as automotive technologies continue to advance. The system can keep up with traffic needs projected for mobility 2030 and beyond.

In conclusion, for freeways as well as county and suburban roads, digital technologies can be employed to raise traffic density and traffic flow velocity. Raising density and velocity during heavy traffic times has the effect that each driver will arrive at his destination sooner, and his vehicle claims its space on freeways and county roads for less time. Thus, the heavy traffic load will be transformed into a much lighter traffic load.

The Synergetic Five-Step Innovation Process (SFSIP) with its Logical-Mathematical Intelligence has proven its powerful impact on creating revolutionary process solutions and for raising productivity enormously.

Epilogue

The 4th industrial revolution is marked by a series of transformations of work processes, products, and services. These transformations have been dominated by fast emerging digital technologies. The opportunities for innovation seem unlimited. The improvements in business and human life surpass previous experience by a wide margin as disruptive change becomes the new norm. Disruptive change is not limited to the end result of an entrepreneur's efforts; rather, it affects the entrepreneur's own work as well. That is actually where the enormous acceleration of change and the multifold impact on productivity originate. The novel Synergetic Five-Step Innovation Process (SFSIP) with the Logical-Mathematical Intelligence (LMI) at its center can be applied as a decisive compass guiding the experienced as well as the novice entrepreneur through the maze of complex conditions toward unknown ideal opportunities.

The combined power of SFSIP and LMI methodologies is unfolded in the analysis of historic innovations and is applied to the demanding needs of modern societies' emerging mass mobility. The disciplined inquiry into disguised aspects of otherwise well-known modernizations discloses the revolutionary trend for the world of innovation. The secret is virtual infrastructure.

The backbones of processes of any kind have been physical infrastructures. Physical infrastructures that enable physical movement and transportation of goods were required throughout the evolution of modern societies and are even more important in today's industrial age. In modern societies, we observe tangible and intangible networks and corresponding infrastructures. As traffic loads on tangible infrastructures such as roadways and bridges increase and exceed the available capacities, traffic collapses. In response, physical expansions have been the norm. Lanes were added to roadways, and bridges were added to the network.

Physical expansions are facing natural space limitations. In some cases, there are natural limitations. For traffic on freeways and county roads, physical expansion is often impossible because of land limitations. Advanced automotive technologies have enabled higher velocities, which have resulted in capacity enhancements, but human comfort causes another natural capacity limitation and pushes the traffic into breakdown.

Electronic technologies are providing a way to avoid the choking effect and open up new avenues. Electronic computing represents the next decisive step into the future. The invention of the Internet, referred to as the "The Information Super Highway," provides a new means of knowledge collection and distribution.

Virtual infrastructure is viewed as the key to continued innovations. Most physical infrastructures seem to have been exploited to their limits. When those limits are reached, system breakdown seems to be the answer. The new thinking in terms of identifying, developing, and implementing a virtual infrastructure that can be superimposed onto the physical infrastructure represents a promising path into the future.

The combined SFSIP and LMI methodologies provide an exceptional guidance for successful journeys in the new space of virtual infrastructures. The aspect of the mathematics involved should not scare the novice entrepreneur. In most cases, the logical-mathematical intelligence does not require more than high school mathematics. We do not have to be an Isaac Newton or an Albert Einstein who worked through exceptionally complex mathematical procedures to arrive at exceptionally simple and transparent correlations that are extremely powerful. Likewise, most natural laws, which can be engaged to derive at logical-mathematical knowledge, are also simple and transparent to someone who is in command of high school level math. The beauty of nature is that it often follows very simple mathematical logic.

The virtual infrastructures created in Chapter 11 for enhancing lane capacities on freeways and county roads, are examples of transparency and simplicity. The Space Division Traffic Management formats for freeways and county roads (SDFTM, SDCTM) illustrate the simplicity of natural physics laws to be employed and the simplicity of the end result. In contrast, the physical implementations of virtual infrastructures, such as the SDFTM and SDCTM formats, are highly complex. The implementation involves high-speed and high-capacity computer systems and an assortment of digital technologies. Once all that has been put together, the final system should be simple again because it has to earn human users' utmost trust and confidence.

Virtual infrastructures are recognized as the gateway to successful industry transformations and extraordinary productivity enhancement in the digital age. Invisible and intangible, they are technology-enabled measures that are superimposed onto traditional physical infrastructures and maximize their capacity without any physical expansion. Thus, they break from the norm and accomplish what seemed to be unthinkable in the recent past while making it possible for ideal conflict-free solutions to become a reality.

References

[1]   Arce, E.B. (2023). A phenomenological study of homestay parents to explore effective leadership styles influencing study abroad student success.

[Unpublished doctoral dissertation]. Alliant International University, 2023

[2]   Bach, Michael, Professor, Dr. rer.nat., Electrophysiologist, University of Freiburg, Germany, Optical Illusions and Similar Vision Phenomena, Errata Kaden Verlag, 2022

[3]   Ballard, Robert, D., The Discovery of the Titanic, Warner Book, Inc. October 1995

[4]   Boorstin, Daniel, J., The Discoveries, A Story of Man's Search to Know his World and Himself, Random House, New York, 1983

[5]   Bristol University (UK), Wilson, E., Physical Field Test of Traffic Jam Dynamics, https://www.youtube.com/watch?v=iHzzSao6ypE , 2015

[6]   Bruneau, Michel, The Blessings of Disaster: The Lessons That Catastrophes Teach Us and Why Our Future Depends on It, Rowman and Littlefield, November 2022

[7]   ChatGPT. (2023, September 19). "Applied AI Execution in Leadership Methods."OpenAI. https://chat.openai.com/c/fcb85482-d73c-4890-908f-7e72d591f2d8. Accessed September 19, 2023.

[8]   Colander, David, C., Economics, Third Edition, McGraw-Hill Company, Inc, New York,1998

[9]   Dennison, K. (2023, March 14) The Impact of Artificial Intelligence On Leadership: How To Leverage AI To Improve Decision-Making. Retrieved September 20, 2023, from https://www.forbes.com/sites/karadennison/2023/03/14/the-impact-of-artificial-intelligence-on-leadership-how-to-leverage-ai-to-improve-decision-making/?sh=355db81a33d9

[10]   Dörner, Diedrich, The Logic of Failure, Metropolitan Books, New York 1996

[11]   Enterprise Architectures, Retrieved September 1, 2023

LeanIX_Whitepaper_EA_Success_Kit_EN.pdf

[12]   Gardner, Howard, Frames of Mind, The Theory of Multiple Intelligences, Basic Books, New York, 1983

[13]   Gartner Report, Top Trends of Data and Analytics, 2023

https://www.qlik.com/us/

[14]   Gleiser, Marcelo, The Island of Knowledge, The Limits of Science and the Search for Meaning, Basic Books, 2014

[15]   Google (2023, September 20). Effective Leadership in the Era of AI. Retrieved September 18, 2023 from

https://www.google.com/search?q=leadership+and+ai&rlz=1C1OPNX_enUS1043US1043&oq=leadership+and+ai&aqs=chrome.0.0i512j0i15i2 2i30j0i22i30l8.11879j0j15&sourceid=chrome&ie=UTF-8

[16]   Gray, Clifford, F., Larson, Erik, W., Project Management, The Managerial Process, McGraw-Hill Higher Education, The McGraw-Hill Company, Inc. New York, 2000

[17]   Gregory, Richard, L., Eye and Brain, The Psychology of Seeing, Fifth Edition, Princeton University Press, New Jersey, 1997

[18]   Hanseatic Trades League, Retrieved July, 2023

https://en.wikipedia.org/wiki/Hanseatic_League

[19]   Hanseatic Cog, Hansa's Flat-Bottom Sailing Vessels, Retrieved July, 2023 https://en.wikipedia.org/wiki/Bremen_cog

[20]   Heisenberg, Werner, Schritte über Grenzen, Gesammelte Reden und Aufsätze, R. Piper & Co. Verlag, München, 1971

[21]   Hisrich, Robert, D, Peters, Michael, P, Shepherd, Dean, A., Entrepreneurship, McGraw-Hill Company, Inc., New York, 2005

[22]   Infinity, Thoughts about Infinity, Retrieved August, 2023

https://en.wikipedia.org/wiki/Infinity

[23]   Isaacson, Walter, Albert Einstein, His Life and Universe, Simon & Schuster, New York, 2007

[24]   Isaacson, Walter, Elon Musk, Simon & Schuster, New York, September 2023

[25]   Isaacson, Walter, Steve Jobs, Simon & Schuster, New York 2011

[26]   Jett, Stephen C., Ancient Ocean Crossings, Reconsidering the Case for Contacts with the Pre-Columbian Americas, University of Alabama Press, Alabama 2017

[27]   Kahn, Kenneth, B., The PDMA Handbook, Product Development & Management Organization, John Wiley & Sons, Inc., New Jersey, 2013

[28]   Kaufman, Scott, B., Gregoire, Carolyn, Wired to Create, Unrevealing the Mysteries of the Creative Mind, Penguin Random House, LLC, 2015

[29]   Larson, Erik, W., Gray, Clifford, F., Project Management, The Managerial Process, Published by Generic, January 2020

[30]   Lewis, Danny, A coal Fire May have helped Sink the Titanic, January 2017 https://www.smithsonianmag.com/smart-news/coal-fire-may-have-helped-sink-titanic-180961699/

[31]   Massachusetts Institute of Technology (MIT), Flynn, M.R., Kasimov, A.R., Nave, J.C., Rosales, R.R., Seibold, B., 2008, Traffic Modeling - Phantom Traffic Jams and Traveling Jamitons - Analytical Simulation of Traffic Jam Formation Dynamics, https://math.mit.edu/projects/traffic/

[32]   Mayer-Schönberger, Viktor, & Cukier, Kenneth, Big data, A Revolution that will Transform how we Live, Work, and Think, Houghton Mifflin Harcourt Publishing Company, New York, 2013

[33]   Molony, Senan, Titanic: Why She Collided, Why She Sank, Why She Should Never Have Sailed, July 1, 2021

[34]   Newton, Isaac, The Principia: Mathematical Principles of Natural Philosophy, Translated by Bernard Cohen and Anne Whitman, Berkley University of California, Press, 1999

[35]   Obama, Barack, A Promised Land, Penguin Random House, LLC 2020

[36]   Paga, P. (2023). Ethics in Artificial Intelligence and its impact on Leadership Styles, [Unpublished doctoral dissertation]. Alliant International University, 2023.

[37]   Patten, Louise, As Good as Gold, The Titanic – What Really Happened, Quercus Publishing, July 2013

[38]   Project Management Institute, A Guide to the Project Management Body of Knowledge (PMBOK) – The Standard of Project Management, Seventh Edition, Published by Project Management Institute, August 2021

[39]   Read, Piers, Paul, Ablaze, The Story of the Heroes and Victims of Chernobyl, Random House, Inc, New York, 1993

[40]   Reinkemeyer, L., Editor, Process Mining in Action, Principles, Use Cases and Outlook, Springer Nature Switzerland AG, 2020

[41]   Samuelson, Paul, A., Nordhaus, William, D., Macro- and Micro-Economics, Fifteenth Edition, McGraw-Hill, Inc. New York, 1948 – 1995

[42]   Schlueter, Georg, HansaTekNetics LLC, *Multi-Lane Traffic Management System for Platoons of Autonomous Vehicles,* Patent Application Filed with the US Patent and Trademark Office (US-PTO), U.S. Serial No.17/730,411, 17-April-2022, Publication on Nov. 2, 2023 (US 2023/0351987 A1)

[43]   Schlueter, Georg, HansaTekNetics LLC, *Traffic Jam Avoidance System that Assigns Vehicles to Lanes Based on Driver Comfort Levels*, Patent NO. 11663909, 20-May-2023

[44]   Schlueter, Georg, HansaTekNetics LLC, *Method of Generating Bi-Directional Green Waves of Traffic by Alternating Lights in Zones*, Patent Application, US PTO, Serial No, 18/328,699, 2-June-2023

[45]   Schlueter, Georg, *Traffic Flow Data Organized into Maximum and Minimum Center-to-Center Intervehicle Spacing for Selected Velocities*, Personal Engineering Handbook, 2014 to Present

[46]   Smith, Adam, The Wealth of Nations, Introduction copyright by Robert Reich 2000, Biographical Note copyright by Random House, Inc. New York, 1994, Original Title: Inquiry into the Nature and Causes of the Wealth of Nations, 1776

[47]   Stair, Ralph, M., Reynolds, George, W., Principles of Information Systems, Twelfth Edition, Cengage Learning, Boston, MA, 2016

[48]   Stanford University, Project Center, Disruptive Technology and Digital Cities (DTDC) Program, https://gpc.stanford.edu/digitalcities

[49]   Steinberg, Mark, Great Minds, Founders of the Scientific Age, Published by We Can't Be Beat LLC, 2016

[50]   Stoob, Heinz, Die Hanse, Styria Verlag, Graz Wien Köln, 1995, The Unique History and Astounding Accomplishments of the Hanse During Medieval Times

[51]   The World Bank: Urban Development, Retrieved April 3, 2023 from:

https://www.Worldbank.org/en/topic/urbandevelopment/overview

[52]   Thiel, Peter, Masters, Blake, Zero to One, Notes on Startups or how to Build the Future, The Random House, LLC, New York, 2014

[53]   Traffic Flow and Highway Capacity, Transportation Research Record, No. 1398, Highway Operations, Capacity, and Traffic Control, National Academy Press, Washington, D.C. 1993

[54]   Traffic Flow Theory, A State-of-the-Art Report, Oak Ridge National Laboratory, Dr. Ajay K. Rathi, Project Lead. 2002, Retrieved from: http://www.tfhrc.gov/its/tft

[55]   Turning Circle of Ocean Vessels, https://www.shipsbusiness.com/turning-circle.html, Retrieved April 3, 2023

[56]   Vance, Ashley, Elon Musk, Tesla, SpaceX, and the quest for a Fantastic Future, HarperCollins Publishers, New York, 2015

[57]   Virtual Infrastructures, https://www.dnsstuff.com/what-is-virtual-infrastructure, Retrieved April 2023

[58]   Wikipedia, Jakob Fogger, https://en.wikipedia.org/wiki/Jakob_Fugger, Retrieved April 2023

[59]   Wikipedia, The Great central Library of Alexandria, Retrieved from the Internet, September 2023, https://en.wikipedia.org/wiki/Library_of_Alexandria

https://www.bing.com/images/search?q=Royal+Library+of+Alexandria&mmreqh=
bbH5tooj1KshBiuGXkhEfN0mqGXL%2bgfJW0PnELHk4dg%3d&form=IDINTS&first=1&cw=1263&ch=581

[60]   World Economic Forum, Migration and its Impact on Cities, Retrieved July 2023

https://www.weforum.org/reports/migration-and-its-impact-on-cities

[61]   Xenon-135 Effect on Nuclear Power Operation, Internet Accessed on November 30, 2023: https://en.wikipedia.org/wiki/Xenon-135

[62]   Batchelor, G. K., An Introduction to Fluid Dynamics, Cambridge Mathematical Library, 2nd Edition

[63]   Chandler, Daniel; Munday, Rod (10 February 2011), "Information technology", A Dictionary of Media and Communication (first ed.), Oxford University Press, ISBN 978-0199568758, retrieved 20 November 2023;

[64]   The Cybersecurity Exchange (n.d.) retrieved from https://www.eccouncil.org/cybersecurity-exchange/threat-intelligence/cyber-kill-chain-seven-steps-cyberattack/

[65]   Merriam-Webster Dictionary, retrieved from https://www.merriam-webster.com/dictionary/cyber

[66]   Greenshield, Bruce, PhD, The Linear Speed-Density Relation, Proc. of the 13th Annual Meeting of the Highway Research Board, Dec. 1933

[67]   Kühne, Reinhart, D., Prof. Dr. German Aerospace Center, Transportation Studies, Berlin Germany, Foundations of Traffic Flow Theory 1: Greenshields' Legacy – Highway Traffic, Recovered from the Internet, January 2024: chromeextension://efaidnbmnnnibpcajpcglclefindmkaj/https://krbalek.cz/For_students/mds/clanky/Greenshields.pdf

www.ingramcontent.com/pod-product-compliance
Lightning Source LLC
Chambersburg PA
CBHW051517120626
46551CB00012B/963